# 18th International Conference Diagnostics of Machines and Vehicles (Diagnostyka 2019)



MATEC Web of Conferences Volume 302 (2019)

Bydgoszcz, Poland
12 December 2019

ISBN: 978-1-5108-9931-5

**Printed from e-media with permission by:**

Curran Associates, Inc.
57 Morehouse Lane
Red Hook, NY 12571

**Some format issues inherent in the e-media version may also appear in this print version.**



For additional information, please contact EDP Sciences – Web of Conferences
at the address below.

EDP Sciences – Web of Conferences
17, Avenue du Hoggar
Parc d'Activité de Courtabœuf
BP 112
F-91944 Les Ulis Cedex A
France

Phone: +33 (0) 1 69 18 75 75
Fax: +33 (0) 1 69 28 84 91

contact-edps@webofconferences.org

**Additional copies of this publication are available from:**

Curran Associates, Inc.
57 Morehouse Lane
Red Hook, NY 12571 USA
Phone: 845-758-0400
Fax: 845-758-2633
Email: curran@proceedings.com
Web: www.proceedings.com

# TABLE OF CONTENTS

## Preface

The scientific conference "Diagnostics of Machines and vehicles", which took place 12 of December in Bydgoszcz, is the continuation of a series of international conferences "Diagnostyka" organized by the Mechanical Engineering Department at the UTP University of Science and Technology. The meeting was focused on the practical application diagnostics theory into industry. Technical diagnostics allow to determine the technical condition of objects, creates the opportunity to achieve significant material and financial savings. Ultimately, it also enables the development of new machine operation strategies, conducting highly economical repair technologies and rational spare parts management.

The main topics of the conference were: progress in testing machines supported by virtual techniques, diagnostic systems and their applications, presentation of research results and their applications, knowledge management about machine operation.

The scientific deliberations were carried out in a plenary session, poster sessions, in which participants of many scientific centres delivered their presentations.

The conference was attended by dozens of participants, including representatives of various companies and universities from Poland and abroad: UTH Technical and Commercial University, Poland; Warsaw University of Life Sciences – SGGW, Poland; UTP University of Science and Technology, Poland; Slovak University of Agriculture in Nitra, Slovak Republic; University of Novi Sad, Serbia; S-EKA spol. s r.o., authorized organization for emission control of road vehicles, Slovak Republic; Institute of Thermomechanics of the CAS, Czech Republic; Lviv Polytechnic National University, Ukraine; Khmelnitsky National University, Ukraine; South Ural State University, Russia; University of Defence Brno, Czech Republic; ARRIVA Slovakia a.s., Slovak Republic; Gdańsk University of Technology, Poland.

We would like to thank all conference participants, authors and reviewers, for their contributions to the success of the conference and already today we would like to invite all participants to the next edition of the conference.

The website of the conference: http://www.dmv.utp.edu.pl.

*Editors*
Michał Liss, Ewa Kuliś, Marcin Łukasiewicz,
Mariusz Żółtowski, Tomasz Kałaczyński

18th International Science Conference
DIAGNOSTYKA

**SCIENTIFIC COMMITTEE:**

| | |
|---|---|
| **Prof. B. Żółtowski** | **Chairman**<br>**UTP Bydgoszcz** |
| Prof. W. Bajkow | AP Mińsk |
| Prof. J. Błachnio | ITWL Warszawa |
| Prof. A. Bochat | UTP Bydgoszcz |
| Prof. L. Castañeda | EAFIT Kolumbia |
| Prof. C. Cempel | Politechnika Poznańska |
| Prof. T. Dąbrowski | WAT Warszawa |
| Prof. S. Gad | Politechnika Świętokrzyska |
| Prof. A. Gajek | Politechnika Krakowska |
| Prof. J. Girtler | Politechnika Gdańska |
| Prof. A. T. Hoppe | Southampton Institute U.K. |
| Prof. A. Jastriebow | Politechnika Świętokrzyska |
| Prof. B. Kolator | UWM Olsztyn |
| Prof. J. Lewitowicz | Politechnika Warszawska |
| Prof. P. Lindstedt | Politechnika Białostocka |
| Prof. M. Matuszewski | UTP Bydgoszcz |
| Prof. S. A. Matyuk | Khmelnitsky National University |
| Prof. J. Musiał | UTP Bydgoszcz |
| Prof. Ł. Muślewski | UTP Bydgoszcz |
| Prof. J. Napiórkowski | UWM Olsztyn |
| Prof. T. Nowakowski | Politechnika Wrocławska |
| Prof. K. Peszyński | UTP Bydgoszcz |
| Prof. S. Radkowski | Politechnika Warszawska |
| Prof. S. Salamon | Politechnika Częstochowska |
| Prof. J. Sempruch | UTP Bydgoszcz |
| Prof. V. Shevchenko | AN Mińsk |
| Prof. P. Simiński | WITPiS Sulejówek |
| Prof. M. Skyba | Khmelnitsky National University |
| Prof. F. Tomaszewski | Politechnika Poznańska |
| Prof. H. Tylicki | PWSZ Piła |
| Prof. S. Wierzbicki | UWM Olsztyn |
| Prof. J. Żurek | ITWL Warszawa |

**ORGANIZING COMMITTEE:**

| | |
|---|---|
| **Dr inż. Tomasz Kałaczyński** | **Chairman** |
| Dr inż. Michał Liss | |
| Dr inż. Marcin Łukasiewicz | |
| Dr inż. Mariusz Żółtowski | |
| Dr inż. Joanna Wilczarska | |
| Dr inż. Piotr Aleksandrowicz | |
| Dr inż. Sylwester Borowski | |
| **Mgr inż. Ewa Kuliś** | **Conference office** |

edp sciences

Web of Conferences

# Statement of Peer review

In submitting conference proceedings to *Web of Conferences*, the editors of the proceedings certify to the Publisher that

1. They adhere to its **Policy on Publishing Integrity** in order to safeguard good scientific practice in publishing.
2. All articles have been subjected to peer review administered by the proceedings editors.
3. Reviews have been conducted by expert referees, who have been requested to provide unbiased and constructive comments aimed, whenever possible, at improving the work.
4. Proceedings editors have taken all reasonable steps to ensure the quality of the materials they publish and their decision to accept or reject a paper for publication has been based only on the merits of the work and the relevance to the journal.

Title, date and place of the conference

Diagnostyka 2019 - Diagnostics of Machines and Vehicles, Bydgoszcz, Poland December 12. 2019

Proceedings editor(s):

Dr hab. inż. Janusz Musiał
Dr hab. inż. Maciej Matuszewski
Dr hab. inż. Łukasz Muślewski
Dr inż. Tomasz Kałaczyński

Date and editor's signature

# Application of impact analysis SDC method for verification of whiplash injuries

*Piotr* Aleksandrowicz[1,*]

[1]Machine Maintenance Department, Institute of the Automation and Transport, University of Science and Technology, Al. prof. S. Kaliskiego 7, 85-796, Bydgoszcz, Poland

**Abstract.** Collisions of vehicles that occur in cities involve relatively low speeds. Even though the speed of a collision is not high passengers are exposed to a risk of neck spine damage when a rear-end accident takes place. These injures are referred to as Whiplash injuries which are difficult to verify when it is necessary to resolve a rear-end collision damage claim. This study presents procedures of SDC method to be used for verification of circumstances that are reported to have caused such injuries. An analysis was carried out based on a case study of a collision of two passenger cars of different categories and the results can be used by theoreticians and practitioners who are involved in research on this subject.

## 1 Introduction

Although development of motorization has certainly a positive impact on the economic growth, and mobility of people, it has also negative effects.

In towns the volume of traffic is increasing which involves congestion and a higher risk of collision. Rush and lack of concentration of drivers contributes to accidents with participation of a vehicle – human – environment (V-H-E). Mutual relations and dependencies between people involved in this process are presented in Fig. 1.

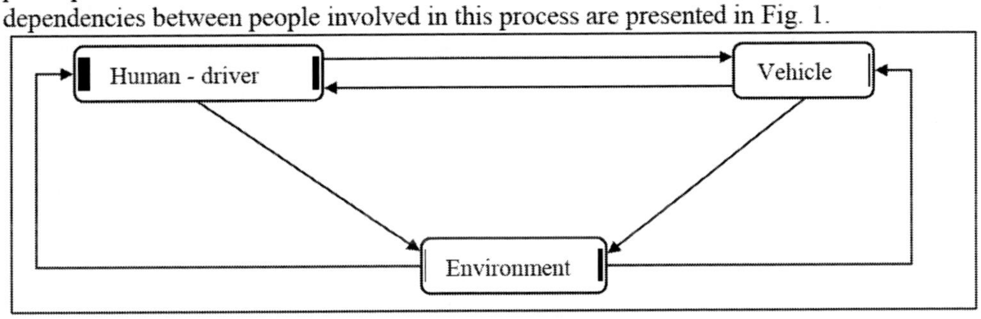


**Fig. 1.** H-V-E system [1].

The most important elements that affect safety of road traffic referred to the H-V-E system include:

- humans – knowledge, skills, concentration, etc.,
- environment – road infrastructure, traffic volume, road signs, etc.,
- vehicle – dimensions, mass, impact resistance, car body deformability and its passive safety devices, etc.

Collisions classified according to a vehicle damage localization and the impact direction are presented in Fig. 2 and those depending on the impact angle in Fig. 3.

* Corresponding author: paleksandrowicz@utp.edu.pl

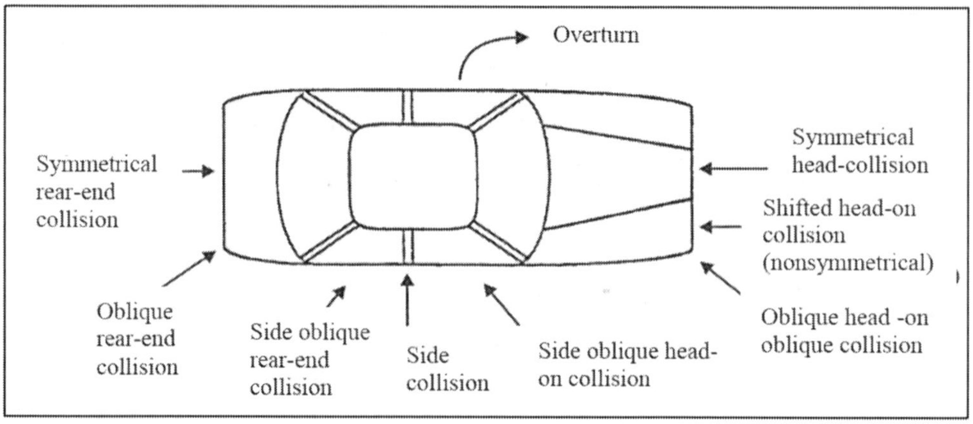


**Fig. 2.** Classification of impacts according to their localization [2].

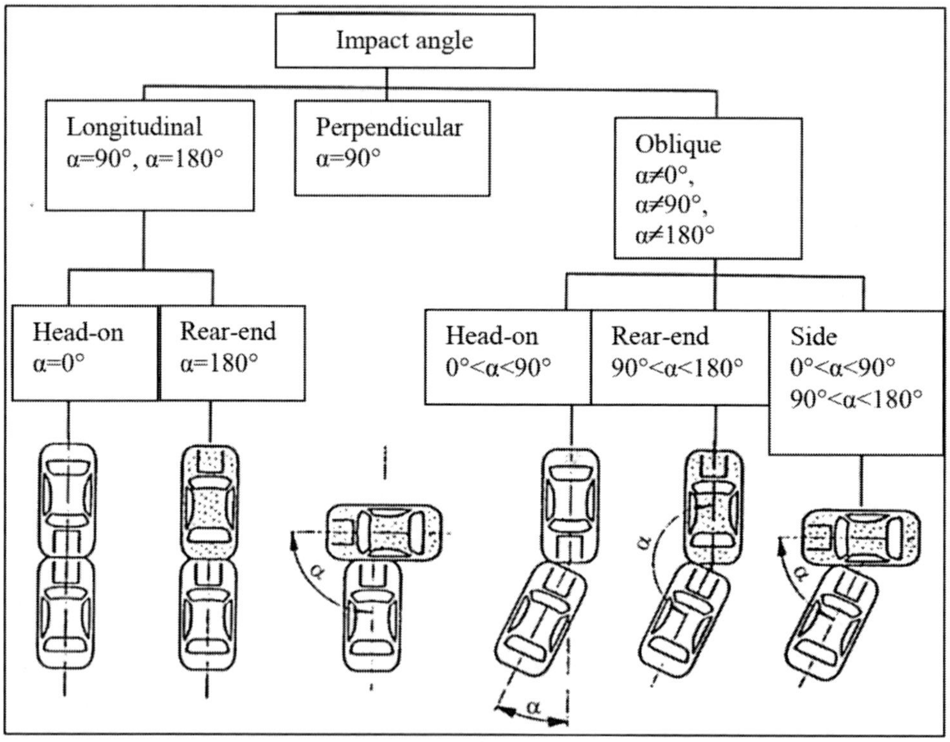


**Fig. 3.** Definition of the impact angle [2].

Head-on collisions are the most dangerous ones, whereas rear-end collisions are considered as the least serious. This results from the speed of the involved vehicles. Therefore, head-on crashes focus more attention than the rear-end ones. For example, work [3] includes an analysis of three different types of bumper beams of vehicles involved in crashes with different obstacles. Based on the obtained results, the authors selected a beam with the best characteristics for impacts involving low collision speeds. However, rear-end accidents, despite relatively low collision speeds, can also cause serious injuries to the neck due to rapid longitudinal forces acting forward from the vehicle back when safety devices fail to provide full protection from these injuries.

This kind of spine injury is referred to as a Whiplash injury. In literature it is also called WAD (Whiplash Associated Disorders), Neck Injury, etc. Fig. 4 shows a mechanism of the above-mentioned injury occurrence.

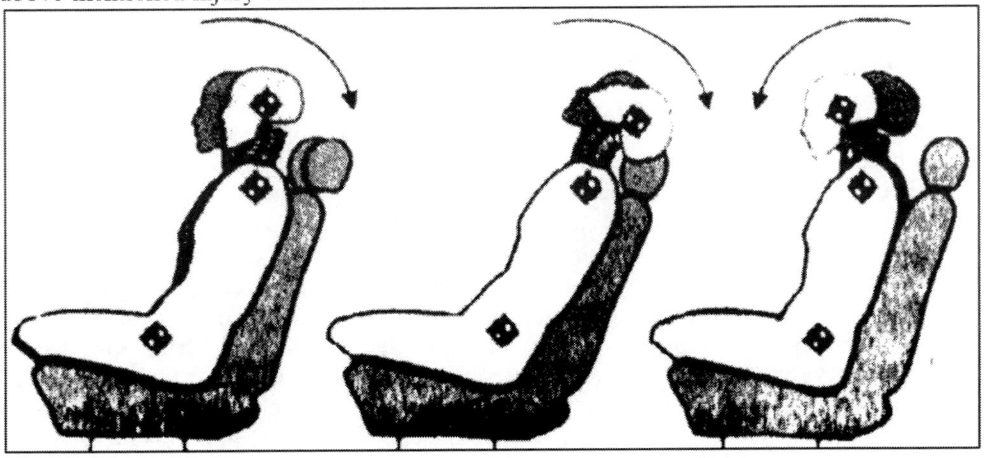

**Fig. 4.** Whiplash injury occurrence mechanism [4].

In practice, it is very difficult to verify claims involving whiplash injuries as it requires application of methods and tools enabling an analysis of many variables and criteria. A limited amount of input data is a significant barrier in verification of traffic claims.

## 2 Research tools and procedures

SDC method (Static, Dynamic, Characteristic analysis) has been proposed to verify an impact and Whiplash injury occurrence probability.

### 2.1 Probability criteria for Whiplash injury occurrence

Mechanism of Whiplash injury occurrence involves a rapid back and forth movement of the neck. This kind of injury causes persistent pain and can also lead to permanent physical disorder. These injuries affect the spine bones and the, so called intervertebral spaces. It was also noticed that chronic pain can occur without backward head movement in the direction to the headrest. In this case the injury can be caused by simultaneous parallel backward movement of the head which also affects the spine. Fig. 5 shows intensity of Whiplash injury in a function of time.

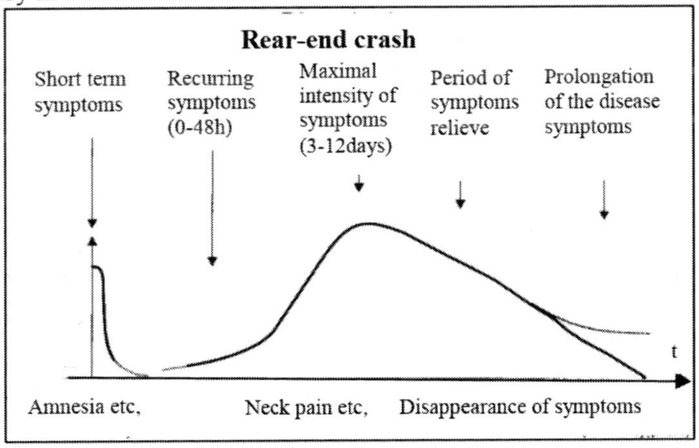


**Fig. 5.** Intensity of Whiplash injury in a function of time [4].

External criteria of Whiplash injury occurrence probability [4]:

- relative speed of the hitting and hit vehicles was higher than 11 $km \cdot h^{-1}$,
- mean acceleration of the hit vehicle exceeded 3 *g* or acceleration of the passenger or driver upper body part inside the vehicle exceeded 6 *g*.

### 2.2 Research procedures of SDC method

Verification of an impact by SDC method requires three research procedures. The first one involves a static analysis of the damage to the vehicles involved in the accident. This procedure is supposed to provide geometric verification of the vehicle damage. The research methods of this procedure are discussed, among others, in work [5], where an analysis of an impact of a truck tractor with a passenger car is presented.

The second procedure uses computer devices for dynamic analysis of road accident reconstruction. It requires solution of problems involved in an impact modeling which are described, for example, in work [6]. This procedure applies simulation programs to perform analyses. Programs which model car accidents in MBS convention (Multi Body System) are most commonly used in practice. This modeling is, however, simplified and an expert needs to change many of the program parameters. For example, in works [7, 8, 9] the authors, after identification of the impact parameters on the basis of a crash test, present sensitivity of the simulation parameters to be used for reconstruction of the crash according to initially accepted input data which causes simplification of modeling the forces that occur during the vehicle collision provided by the simulation programs. In Poland the most popular program used for impact modeling in MBS convention is V-SIM4 program. It is also used for simulation tests of, among others, the influence of selected defects of the braking system on the possibility to stop the car and on the impact effects which is described in works [10, 11]. These works provide a description of modeling the movement of a vehicle with defects of a braking system with the use of V-SIM4 program in dependence on the brake shoes wear degree.

The third research procedure involves an analysis of characteristic damage in the site of the objects contact which was caused during the impact, for example pieces of paint, traces of wheels, etc. Occurrence of this kind of marks is accompanied by complicated phenomena which are described and analyzed, among others, in work [12]. This work includes information about how such marks occur during an impact of cars and provides the possibilities of practical applications of the results of carried out research.

## 3 Research objects and parameters of the movement environment

Research objects were passenger cars Mercedes class E of third generation and Citroën C1. The most important technical data of the cars and parameters of the movement environment accepted for simulation calculations are presented in table 1.

**Table 1.** Technical data of research objects and parameters of the movement environment.

| **Parametr** | **Citroën** | **Mercedes** |
|---|---|---|
| Vehicle curb weight | 875 kg | 1 351 kg |
| Driver weight | 75 kg | 75 kg |
| Vehicle length/ width | 3.435 m /1.630 m /1.470 m | 4.740 m/1.740 m/1.430 m |
| Wheelbase | 2.340 m | 2.800 m |

| | | |
|---|---|---|
| Vehicle brake efficiency | 8.2 kN | 16.3 kN |
| ABS | YES | No |
| Dry asphalt road | adhesive friction coefficient $\mu_1$=0.8 slip friction coefficient $\mu_2$=0.75 | adhesive friction coefficient $\mu_1$=0.8 slip friction coefficient $\mu_2$=0.75 |
| Rolling resistance coefficient | 0.015 | 0.015 |

## 4 Case study – own research

The verification was performed according to SDC method. Vector silhouettes of the vehicles were used for comparison of geometric location of the vehicles, Fig. 6. They are in the same scale and provide accurate shapes of the vehicle bodies (https://www.autoview.at). Results of measurements carried out according to the vector silhouettes confirmed overlapping of the vehicle damage zones.

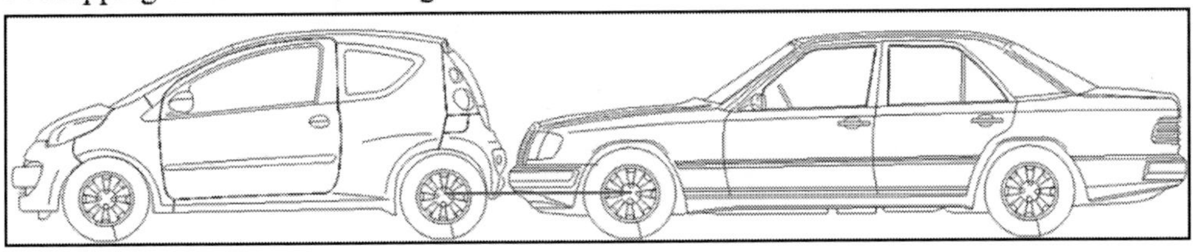

**Fig. 6.** Comparison of the vehicle bodies.

Energy required to cause deformation of the body of Citroën was estimated on the basis of Equivalent Energy Speed from EES catalogue (http://www.dsd.at) and it was 20 km·h$^{-1}$. Dynamic calculations showed that inertial accelerations reached the value of about 6.66 *g*. The impact scheme is demonstrated in Fig. 7.

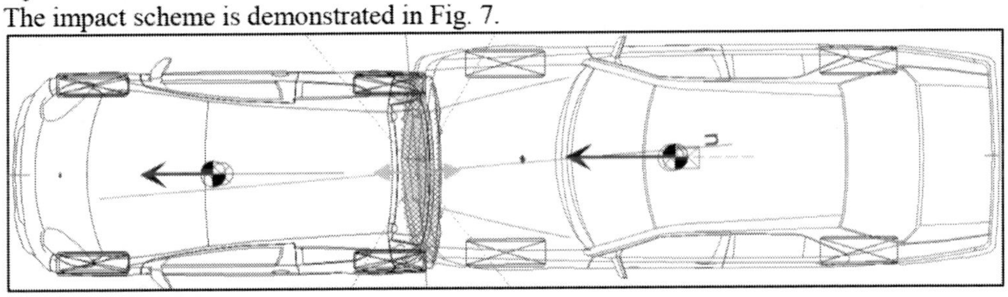

**Fig. 7.** Collision of Citroën and Mercedes passenger cars.

Below, in Fig. 8, there are time courses of acceleration in axes of a system connected with the vehicle x, y, z, and in Fig. 9 there are inertial acceleration changes that affected the driver of Citroën during an accident every 20 ms – they are represented on the below schemes by red vectors.

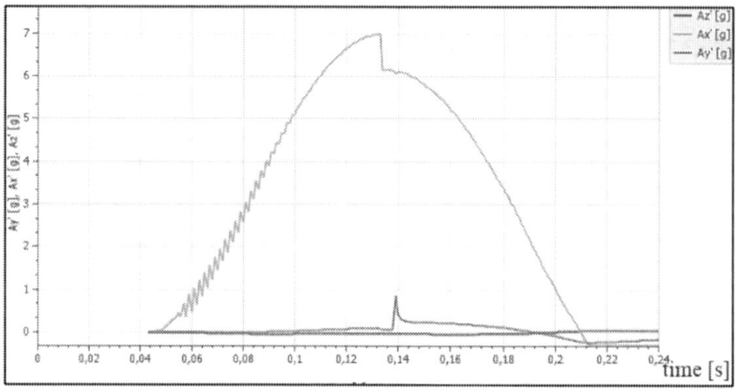


**Fig. 7.** Change in inertial accelerations in a function of time.

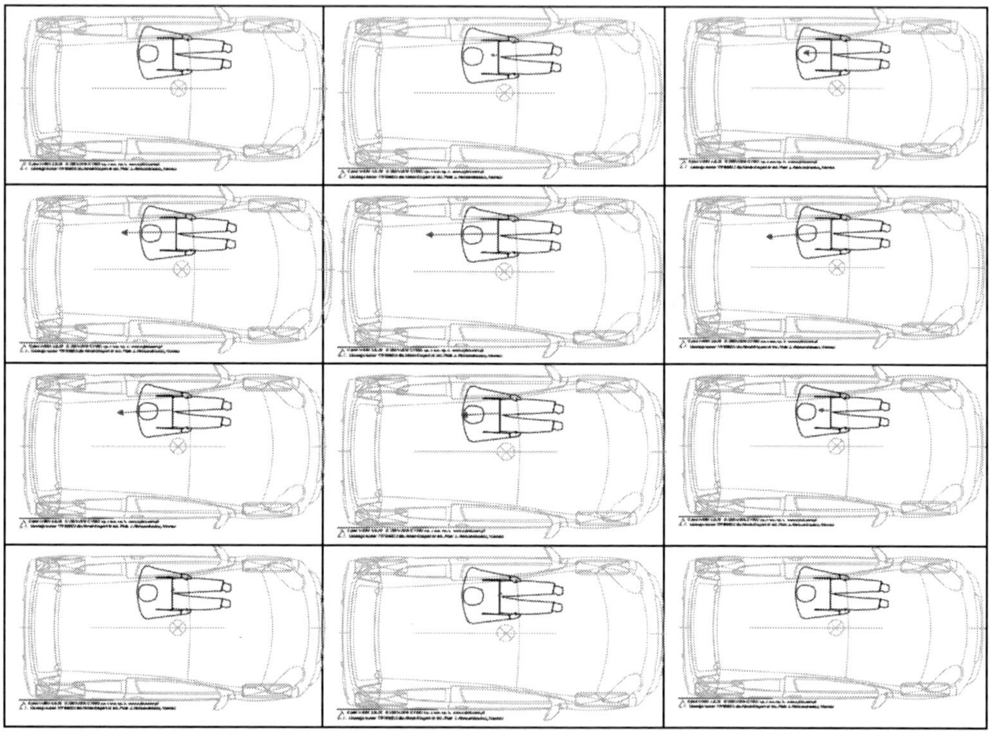

**Fig. 8.** Direction changes in inertial accelerations.

# 5 Conclusions

The tests have revealed the following:

- the body of the driver of Citroën was affected by inertial accelerations in a very short time, from the impact until culmination is 80 ms, reaching the value of about 6.66 *g* and in backward direction– hence, reaching a value higher than the boundary one -3 *g*,
- the criteria that need to occur to cause Whiplash injury to the driver of Citroën were met,
- analysis of the impact according to SDC convention makes it possible to efficiently verify consistence of the reported damage was with occurrence of the mechanism that causes Whiplash type of injury.

## References

1. J. Wicher. *Motor vehicles. Car safety and road traffic.* WKŁ Warsaw. pp. 125-126, (2004)
2. J. Wicher. *Assessement of road traffic, Classification of car collisions.* SRTSiRD Warsaw. pp. 142-143, (2006)
3. G. Gulyaev, V. Loginov, N. Kozlov. *Method of Designing the Superstructure of the Car Body Based on the Requirements of Low-Speed Collisions.* 9th International Scientific Practical Conference on Innovative Technologies in Engineering, Journal of Physics, Conference Series Vol. 1059, Russia, pp. 140-147, (2018)
4. L. Prochowski, J. Unarski, W. Wach, J. Wicher. *Automotive vehicles. Basics of road traffic accidents.* WKŁ Warsaw. pp. 244-245, (2008)
5. P. Aleksandrowicz. *Verifying a truck collision applying the SDC method.* 58th International Conference of Machine Design Departments, 6-8.09.2017, Praque, Czech Republic, pp.14-19, (2017)
6. J. Zalewski. *Selected problems of motor vehicle maintenance after side impact collision.* MATEC Web Conf. Paper No. 01019 182, (2018)
7. R. Kostek, P. Aleksandrowicz. *Simulation of car collision with an impact block.* 11th International Congress of Automotive and Transport Engineering, 8-10.11.2017 Pitesti, Romania, (2017)
8. R. Kostek, P. Aleksandrowicz. *Simulation of the right-angle car collision based on identified parameters.* 11th International Congress of Automotive and Transport Engineering, 8-10.11.2017 Pitesti, Romania, (2017)
9. P. Aleksandrowicz, R. Kostek. *Study of vehicle crashes into a rigid barrier.* Transactions of the Canadian Society for Mechanical Engineering, (2019)
10. P. Aleksandrowicz. *The impact of a vehicle braking system state on safe driving – part one.* 17th International Conference Diagnostic of Machines and Vehicles, 19-20.09.2018, Pieczyska, Poland, MATEC Web Conf. vol. 182, 01025, (2018)
11. P. Aleksandrowicz. *The impact of a vehicle braking system state on safe driving – part two.* 10th International Scientific Session Applied Mechanics, 23.11.2018, Bydgoszcz, Poland, AIP Conference Proceedings 2-77, 020001, (2019)
12. K. Brösdorf, M. Görtz. *Gefahren oder gestanden?* Tagungsband des 23rd EVU Jahrestagung. pp.1-6, (2014)

MATEC Web of Conferences **302**, 01002 (2019) https://doi.org/10.1051/matecconf/201930201002
*18th International Conference Diagnostics of Machines and Vehicles*

# Ground vibrations caused by wind power plant work as environmental pollution - case study

*Sylwester* Borowski[1,*]

[1]UTP University of Science and Technology, Al. prof. S. Kaliskiego 7, 85-796 Bydgoszcz, Poland

**Abstract.** The paper presents issues related to the impact of wind farms on the environment. Emphasis was placed on vibrations that are transferred to the ground through the foundations. As research has shown - a case study - vibrations are felt up to about 1000 m from wind farms. According to other literature sources, this may affect living organisms in the ground.

## 1 Introduction

The increasing demand for civilization for energy requires the construction of new power plants supplying eclectic energy to even the most distant places. At the same time, the growing threat of environmental pollution resulting from exhaust emissions, which also includes greenhouse gases, has increased interest in renewable energy sources. One of the ways to generate electricity by this way is to build wind farms [1, 2, 3].

In Polish conditions, Horizontal Axis Wind Turbines (HAWT) predominate in commercial applications. Such a turbine consists of a nacelle with a rotor placed on the tower. A rotor is used to convert wind energy into mechanical energy in the form of a rotary motion. The rotor is connected to a generator that converts rotary motion into electricity.

A typical synchronous generator produces electricity at speeds above 1500 rpm. Due to the large rotor diameters (e.g. Vestas V112 3 MW - 112 m), they must rotate at speeds resulting from the critical speeds of the blade ends. For this size of wind farms, the rotor speed is 15 (to 20) rpm. To obtain the right rotational speed of the generator, gears increasing the number of rotations are used in these constructions. The introduction of this structural element reduces the efficiency of energy conversion and increases the number of elements, which entails greater failure rate and the generation of additional waste (lubricants, oils). Gears and a high generator speed are also responsible for the higher noise and vibration emissions caused by the turbine. These types of turbines are called gear drive turbines.

Gearless turbines generate electricity in low-speed, multi-pole, synchronous ring generators. The rotor in this type of construction is connected directly to the generator, which according to the manufacturer causes an increase in electricity production. The advantages of this type of power plant include, apart from reducing the number of rotating elements (susceptible to wear), also reducing the sources of noise and vibration. The advantages of such a solution were noticed on the difficult German market, where the percentage share of the installed capacity of gearless turbines for this manufacturer is over 50% and increases. Other wind turbine manufacturers also offer such structures (GoldWind 1.5MW, Leitwind 1.5 to 1.8MW, Mtorres 1.65MW, XEMC 2MW, GE / ScanWind 3.5MW).

HAWT wind turbine rotors are most often built as three-blade, two or single-blade rotors found in older solutions are also found. Propeller blades are made of fiberglass or carbon reinforced with polyester. The use of carbon fibers allows you to increase the size of

* Corresponding author: sylwa@utp.edu.pl

the panels, which allows you to obtain greater power. The gondola has the ability to rotate 360 degrees so that it is always in the optimal position during work. This is possible by using a motor together with a gearbox to determine the position of the nacelle in relation to the wind direction. This information is obtained by top-mounted measuring instruments. The control system also uses this information to determine the blade pitch and generator load. In addition, the nacelle includes: a transformer, bearings, lubrication systems and a brake ensuring stopping the rotor in an emergency.

Wind power mast can be built as a closed pipe or truss. A truss tower requires less material at the same strength. However, the closed tower in the form of a pipe, according to popular opinion, looks more aesthetically pleasing and allows safe placement inside the apparatus necessary for the operation of the gym. A tower with a height equal to the diameter of the rotor is considered the most proportional and this is the standard height adopted by most manufacturers. Steel towers, 20-30 square meters in diameter, are delivered in parts to the construction site. The towers have a conical shape, with a diameter increasing towards the base. This shape ensures high strength and material saving. The prefabricated segments are assembled together at the gym assembly site. The 50 m high tower weighs approx. 40 tons.

The wind power plant is founded on a previously prepared foundation. Foundations for wind turbine towers are most often built as reinforced slab foundations.

In difficult areas or in order to save the amount of concrete intended for the foundation, prestressed foundation wells (PFW) and prestressed slab-pile foundations (PPF) are shown in Figure 1. The PFW foundation, otherwise a prestressed foundation well, consists of a reinforced concrete cylinder of the large diameter of row 4 -5 m and a length of 7.5-10.5 m. The cylinder is terminated on the underside of a reinforced concrete slab with a thickness of 0.90 m, while from the top a slab with a thickness of about 0.30 m. Inside the well is placed unpolluted soil from the excavation of the foundation space. The depth of PFW foundation depends on the applied loads coming from the turbine and the surrounding soil at the target location. The PPF foundation is the reinforcement of wind towers in the form of prestressed slab and pile foundations. The PPF foundation consists of a reinforced concrete slab with a diameter of approx. 7.5 m, a depth of approx. 1.5 m supported usually by 12-20 pieces, 10-15 m long piles distributed on a circle with a diameter of 6.0 m. Turbines with greater power than 2 MW may require more piles arranged in two rows [4].

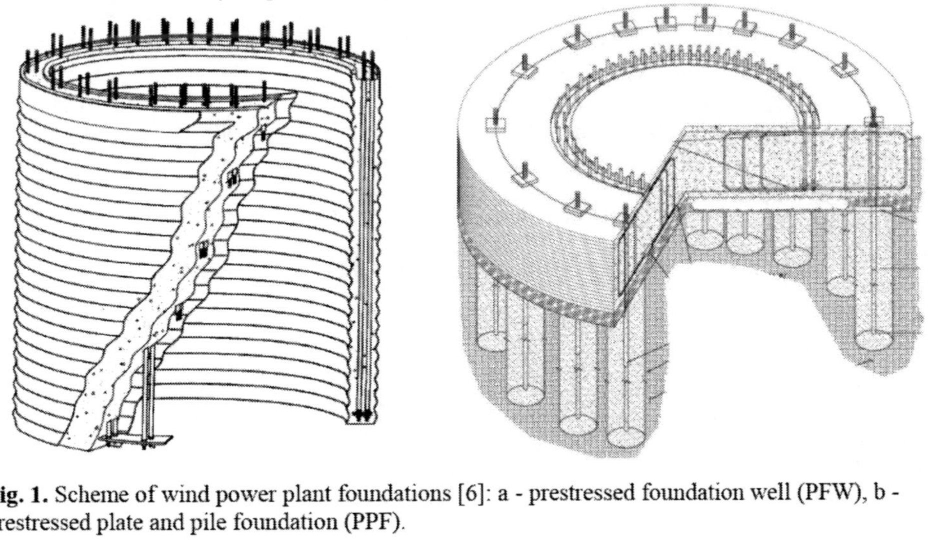

**Fig. 1.** Scheme of wind power plant foundations [6]: a - prestressed foundation well (PFW), b - prestressed plate and pile foundation (PPF).

As it results from the presented short characteristics, a wind power plant is a complicated system consisting of a wind turbine connected to a generator, together with all types of systems necessary for its operation. During its use, the gym is exposed to variable loads resulting from gusts of wind affecting both the rotor blades, as well as the tower and nacelle. These interactions are transferred to the foundation slab and further to the ground. Also, significant blade tip speeds and rotor operation in the air jet are a source of noise. The centrifugation of huge masses, which consists of the mass of the rotor, the mass of wheels and shafts as well as bearings, as well as the rotor of the generator is a source of vibration and noise. Even at home, humans and animals are exposed to wind farms. They are primarily infrasound that can penetrate through concrete partitions.

## 2 Vibrations and their impact on living organisms

Mechanical vibrations generated in the wind power plant are transferred through the foundation to the surrounding ground. Vibration can be generated as a result of normal operation and may increase as a result of damage [5]. A person exposed to this factor is exposed to damage to tissues and blood vessels. Prolonged exposure to vibration can cause lesions in the vascular, osteoarticular and nervous systems.

Changes in the osteoarticular system of the hand arise mainly as a result of local vibrations with frequencies less than 30 Hz. Observed include deformation of joint gaps, calcification of the joint capsules, periosteal changes, changes in bone tissue.

Disorders in the functioning of internal organs are mainly the result of stimulation of individual organs to resonance vibrations (the natural frequencies of most organs are in the range of 2 ÷ 18 Hz).

The functional effects caused by vibrations are [6]:

- increased motor reaction time,
- increased visual response time,
- disruption of coordination of movements,
- excessive fatigue
- insomnia
- irritability
- memory impairment.

As a result of research conducted in our own research team, the effects of noise, infrasound and vibration were found near the wind farm [7, 8].

## 3 Research methodology

The wind farm was the object of research carried out in Rypin - Rypałki. The wind turbine is located at 53.109670 north latitude, 19.434430 east longitude. The tested wind turbine was produced by Vestas. The installed V90 2 MW model has the following parameters [9]:

**Rotor**

Diameter: 90 m

Sweep area: 6.362 m2

Nominal turnover: 14.9 rpm

Rotation range: 9.0-14.9 rpm

Number of blades: 3

**Mast**

Hub height: 95 m

**Weight**

Gondola: 68 t

Rotor: 38 t

Tower: 200 t

The aim of the study was to assess the propagation state of ground vibrations generated by the wind power plant at a distance of up to 940 meters from the wind power plant mast. Preliminary tests were carried out to assess the level of vibrations emitted by the tested wind power plant. The tests were carried out using the characteristics of the vibration signal, which is a good carrier of information about the technical condition of the object.

The measurements of the characteristics of the vibration signal on the test stand were made using a single-axis piezoelectric sensor. This sensor was connected using shielded cables to the four-channel VIBDAQ+ data activation module.

The machine was directly picked up by a piezoelectric PCB Piezotronics sensor ICP model 352C68. The photo of the vibration sensor used during the experimental tests.

At first, vibration measurements were made on the wind power plant mast. Then measurements were taken in one direction from the wind farm, every 50 m. Due to the large number of results obtained, selected vibration signal reception points will be discussed.

## 3 Research results and discussion

The effective value of the vibration process signal was analysed (Fig. 2). The selected measure is one of the most commonly used to describe and quickly assess the level of vibration propagation in vibration diagnostics. It is used in asymmetrical vibrations and takes into account the average value in time domain [1].

The measurements were made for a wind speed of 3.5 m s-1 at a height of 2 m above ground level.

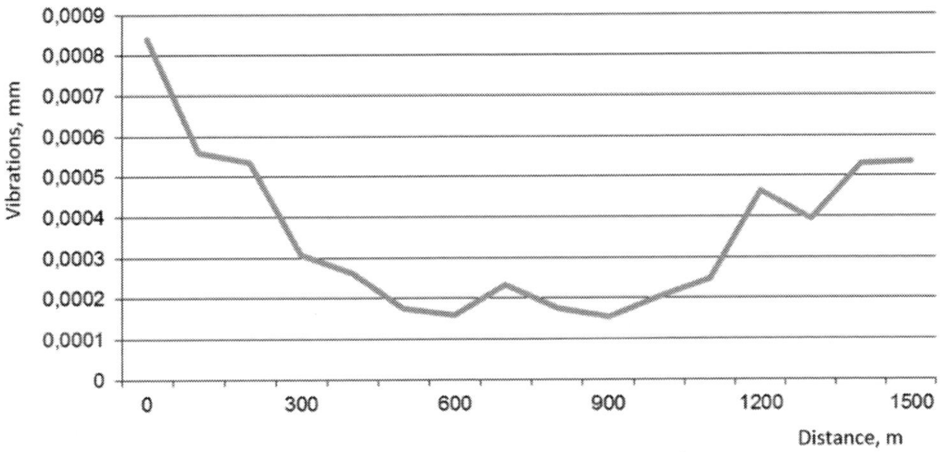


**Fig. 2.** Graph of vibration effective value depending on the distance from the turbine.

The tests are of exploratory nature, in order to increase the accuracy of the diagnosis it would be necessary to make measurements in subsequent time intervals. They indicate the need for further verification of soil vibration propagation in the wind power plant operation process.

The analysis of the obtained test results indicates that the wind power plant generates low frequency vibrations up to 1 Hz. In connection with the above fact, that the level of generated vibration signal is lower than 1 Hz, it should be concluded that it is not harmful to the human body and is safe for surrounding.

There is a need to verify the propagation of soil vibrations, because for points located on the same substrate this correlation is maintained and I indicate the damping of vibrations, while the change of soil and soil properties do not give such an unequivocal conclusion.

## References

1. C. Cempel, *Diagnostyka Wibroakustyczna Maszyn*, 29.08.2011, http://www.wbc.poznan.pl/dlibra/docmetadata?id=8900&from=publication (1998)
2. W. M. Lewandowski, *Proekologiczne odnawialne źródła* (2010)
3. Z. Lubośny, *Farmy wiatrowe w systemie elektroenergetycznym*, (2009)
4. CIOP, http://www.ciop.pl/6538.html (2011)
5. S. Borowski, M. Szubartowski, L. Knopik, K. Migawa, MATEC Web of Conferences 182, 01015 (2018)
6. Fundamenty, http://polishwindenergy.com/index.php/pl/news/144-fundamenty-redukcja-kosztow-o-30-przy-zastosowaniu-technologii-ssf-i-spp (2011)
7. M. Karwowska, J. Mikołajczak, S. Borowski, Z. J. Dolatowski, J. Marć-Pieńkowska, W. Budziński, Ann. Anim. Sci., **14** (2014)
8. M. Karwowska, J. Mikołajczak, Z. J. Dolatowski, S. Borowski, Ann. Anim. Sci., **15** (2015)
9. V90, http://www.vestas.com/Files%2FFiler%2FPL%2FBrochures%2FV90_2_PL.pdf (2011)

# The method of assessment and formation the availability of the subsystem ensuring fitness for means of transport

*Sylwester* Borowski[1,*], *Anna* Mazurkiewicz[1], and *Magda* Czyżewska[1]

[1]University of Science and Technology, Faculty of Mechanical Engineering, 85-796 Bydgoszcz, Poland

**Abstract.** In systems of means of transport operation in order to achieve appropriate completion of assigned transportation tasks it is necessary to maintain a required number of means of transport in the state of availability for carrying out of transportation task (roadworthy and stocked). In general, the processes of rendering vehicles roadworthy are connected to supplying them with fuel and operational materials, carrying out services and repairs, condition diagnostics. In the analyzed system of transport means operation, the processes are carried out in serviceability assurance subsystem SAS. In complex operation systems, the processes of rendering technical objects roadworthy are carried out at specifically designed technical infrastructure posts. The possibility of carrying out the assigned service and repair tasks depends on the availability and the number of such posts. The article presents the method of defining the operational availability and the number of technical infrastructure posts required for appropriate functioning of assigned service and repair task. Then typical calculation results are presented in charts prepared on the basis of data obtained from tests at existing transport means operation system.

## 1 Introduction

In systems of means of transport operation in order to achieve appropriate completion of assigned transportation tasks it is necessary to maintain a required number of means of transport in the state of availability for carrying out of transportation task (roadworthy and stocked). While carrying out the transportation tasks as a result of coercive factors the exhaustion of applied potential of the means of transport takes place (they undergo damage and/or stocking is necessary). From the point of view of the efficiency of the performance of transport system, unavailable vehicles should be rendered roadworthy at a possibly shortest time. As a result, most transport systems makes use of its own technical backup area equipped with an appropriate number of stationary service and repair posts as well as mobile units. In general, the processes of rendering vehicles roadworthy are connected to supplying them with fuel and operational materials, carrying out services and repairs, condition diagnostics, transporting damaged technical objects, rearming in order to complete the assigned task. In the analyzed system of transport means operation, the processes are carried out in separate subsystem. The subsystems of the type may function properly only when appropriate availability and number of service and repair posts is granted.

The problem of control of the processes carried out at serviceability assurance subsystems from the point of view of evaluation criteria such as reliability and availability is discussed in numerous scientific publications [1, 4]. It reflects issues connected to

* Corresponding author: sylwester.borowski@utp.edu.pl

selection of optimal strategy (policy) of servicing and repair as well as evaluation of the operation of serviceability assurance subsystems (service and repair posts). The authors of papers [6, 11] suggested that the methods of shaping the availability of technical backup area posts, whereas papers [9, 10] discuss the methods of the evaluation of the effectiveness and productivity of processes carried out at service and repair posts. Paper [3] discusses methods of defining the structure which may link posts at vehicle service systems, whereas the way of defining technically valid number of service and repair posts in transport systems was described in paper [8]. Among the many methods supporting the evaluation and control process, the semi-Markov decisive processes have been implemented [2], non-deterministic methods of defining optimal solutions (genetic algorithms, evolutionary algorithms, Monte Carlo method) [5, 7], as well as methods and models of mass service theory [12]. The objective of this paper is to work out a method of determining the number and availability of serviceability assurance subsystem post at such a level that would ensure appropriate carrying out of service and repair tasks assigned for to be carried out at such posts.

## 2 Operational availability of serviceability assurance subsystem posts

In systems of means of transport operation in order to achieve appropriate completion of assigned transportation tasks it is necessary to maintain a required number of means of transport in the state of availability for carrying out of transportation task (roadworthy and stocked). In general, the processes of rendering vehicles roadworthy are connected to supplying them with fuel and operational materials, carrying out services and repairs, condition diagnostics. In the analyzed system of transport means operation, the processes are carried out in serviceability assurance subsystem SAS. The subsystem of the type may function properly only when appropriate operational availability of service and repair posts is granted. Presented below are formulas defining operational availability of individual post of serviceability assurance subsystem.

If

$$V_{ij}(t) = P(X_{ij} < t),\ i = 1, 2, ..., p,\ j = 1, 2, ..., q_i \tag{1}$$

is the distribution function of serviceability time $X_{ij}$ of a single post at serviceability assurance subsystem, while

$$W_{ij}(t) = P(Y_{ij} < t),\ i = 1, 2, ..., p,\ j = 1, 2, ..., q_i \tag{2}$$

is the distribution function of renovation time $Y_{ij}$ of a single post at serviceability assurance subsystem (where: $p$ – number of posts groups, $q_i$ – number of posts of $i$-th group), then the operational availability of a single post of serviceability assurance subsystem, determined as probability that over time interval $<t, t+\tau)$ the post is available and provided for, is defined by the formula

$$A_{O_{ij}}(t,\tau) = R_{ij}(t+\tau) + \int_0^t R_{ij}(t+\tau-x)\, dH_{ij}(x) \tag{3}$$

where:

$R_{ij}(t)$ – the function of reliability of a single post of the serviceability assurance subsystem,
$H_{ij}(t)$ – the function of renovation of a single post of serviceability assurance subsystem.

In the case, in which $t \to \infty$, the function (3) has limit value

$$A_{O_{ij}}(\tau) = \lim_{t\to\infty} A_{O_{ij}}(t,\tau) = \frac{1}{E(X_{ij}) + E(Y_{ij})} \cdot \int_\tau^\infty R_{ij}(x)\, dx \tag{4}$$

where:

$E(X_{ij})$ – the expected value of the time of serviceability of a single post of the serviceability assurance subsystem,

$E(Y_{ij})$ – the expected value of the time of renovation of a single post of the serviceability assurance subsystem.

## 3 Operational availability of serviceability assurance subsystem

Operational availability of serviceability assurance subsystem depends on the structure coupling the individual posts or groups of such posts. In the case of the existing system of means of transport operation the individual groups of serviceability assurance posts may be coupled in various ways. Complexity of the existing structure coupling the individual post groups depends primarily on the equipment of posts, worker qualification as well as the type of tasks to be carried out at individual post groups.

### 3.1 Operational availability of posts group of serviceability assurance subsystem with a threshold structure

Presented below are formulas defining operational availability of individual post group of a serviceability assurance subsystem in the case of the posts of the analyzed system are coupled via a threshold structure. Operational availability of $i$-th posts group consisting of $j = 1, 2, ..., q_i$ homogenous posts coupled by threshold structure „$k_i$ with $q_i$" defined by formula

$$A_{O_i}(t,\tau)=\sum_{c=k_i}^{q_i}\binom{q_i}{c}\cdot\left[\overline{A_{O_{ij}}(t,\tau)}\right]^{c}\cdot\left[1-\overline{A_{O_{ij}}(t,\tau)}\right]^{q_i-c} \tag{5}$$

In the case, in which $t \to \infty$, the function (5) has limit value

$$A_{O_i}(\tau)=\sum_{c=k_i}^{q_i}\binom{q_i}{c}\cdot\left[\frac{1}{\left[E\left(\overline{X_{ij}}\right)+E\left(\overline{Y_{ij}}\right)\right]}\cdot\int_{\tau}^{\infty}\overline{R_{ij}(x)}\,dx\right]^{c}\cdot\left[1-\frac{1}{\left[E\left(\overline{X_{ij}}\right)+E\left(\overline{Y_{ij}}\right)\right]}\cdot\int_{\tau}^{\infty}\overline{R_{ij}(x)}\,dx\right]^{q_i-c} \tag{6}$$

### 3.2 Operational availability of serviceability assurance subsystem with a serial structure

The general instance is connected to the situation involving individual serviceability assurance subsystem groups, due to specialist equipment as well as worker qualifications may not replace each other and are supposed to carry out tasks of various type and range. Thus, the post groups of serviceability assurance subsystem are coupled by serial structure. Then, operational availability of serviceability assurance subsystem is determined as product of the availability of its individual groups:

– over time interval $\langle t, t+\tau)$

$$A_O(t,\tau)=\prod_{i=1}^{p}A_{O_i}(t,\tau) \tag{7}$$

– in the case, in which $t \to \infty$

$$A_O(\tau)=\prod_{i=1}^{p}A_{O_i}(\tau) \tag{8}$$

## 4 Operational availability of the serviceability assurance subsystem in carrying out the assigned task

Operational availability of the serviceability assurance subsystem is understood as capability to carry out the assigned service and repair task. Each task assigned to SAS is determined by the length of the time interval $\tau$ devoted to the completion of the task, the size of the task (how many technical objects should be rendered roadworthy and/or stocked) as well as the scope of the task (what should be done). The measure of operational availability of SAS posts in carrying out of the assigned task is the product of the probabilities of two events taking place:

– the event of SAS posts being available at any point (roadworthy and stocked) to undertake the assigned task and will remain in this condition for the time interval $\tau$ of the task duration; this probability is expressed through the value of operational availability of the analyzed subsystem,
– the event of the task assigned at the posts of the subsystem in question will be carried out, i.e. in the time interval $\tau$ the number of objects rendered technically roadworthy will be higher than $k$; this probability is expressed through the value of the productivity index of SAS posts (the way of defining the productivity index is presented in paper [10]).

Taking the above discussion into consideration, the operational availability of the serviceability assurance subsystem in carrying out the assigned service and repair task in the time interval $\tau$, is defined as follows

$$A_{OZ}^{(k)}(\tau)=A_O(\tau)\cdot Z^{(k)}(\tau) \tag{9}$$

where:
$A_O(\tau)$ – operational availability of the posts of SAS defined as the probability of this subsystem being available at any point $t$ and remaining in this condition over a required time interval $\tau$,
$Z^{(k)}(\tau)$ – probability of the number of technical objects rendered roadworthy at SAS posts in the time interval $\tau$ being bigger than $k$.

## 5 Required availability of the serviceability assurance subsystem in carrying out the assigned task

Required operational availability of the SAS for carrying out of service and repair task (rendering $k$ number of technical objects roadworthy in the time interval of $\tau$ length), determined as product of required availability $A_{O_{req}}^{(k)}(\tau)$ as well as the required value of productivity index $Z_{req}^{(k)}(\tau)$ for analyzed posts in the time interval of $\tau$ length is realized in the formula

$$A_{OZ_{req}}^{(k)}(\tau)=A_{O_{req}}^{(k)}(\tau)\cdot Z_{req}^{(k)}(\tau) \tag{10}$$

Service and repair task assigned to the SAS is determined by the required number $k$ of technical objects which should be rendered roadworthy and/or stocked in the given time interval of $\tau$ length, at posts for this subsystem. Whereas the required availability $A_{O_{req}}^{(k)}(\tau)$ of SAS determined for number $k$ of technical objects rendered roadworthy in the time interval of $\tau$ length is described as follows

$$A_{O_{req}}^{(k)}(\tau)=\frac{T_{req}^{(k)}(\tau)}{T_{req}^{(k)}(\tau)+U_{req}^{(k)}(\tau)} \tag{11}$$

where:
$T_{req}^{(k)}(\tau)$ – required availability time for SAS posts for given $k$ and $\tau$,
$U_{req}^{(k)}(\tau)$ – required unavailability time for SAS posts for given $k$ and $\tau$.

Assuming that for any time interval of $\tau$ length the sum of required availability and unavailability times for SAS equals the time interval of $\tau$ length, i.e. $T_{req}^{(k)}(\tau)+U_{req}^{(k)}(\tau)=\tau$, then formula (11) may be written as follows

$$A_{O_{req}}^{(k)}(\tau)=\frac{T_{req}^{(k)}(\tau)}{\tau} \tag{12}$$

Required availability time $T_{req}^{(k)}(\tau)$ for SAS posts depends on the anticipated total time of rendering $k$ number of technical objects (transport means) roadworthy in the time interval of $\tau$ length as well as the $q$ number of uniform posts of the analyzed subsystem (parallel channels for rendering roadworthiness) and is described by the condition

$$T_{req}^{(k)}(\tau)=\frac{k\cdot\overline{U^{OT}(\tau)}}{q}\leq\tau \tag{13}$$

where

$\overline{U^{OT}(\tau)}$ – mean time of technical object remaining at SAS (mean time of rendering technical objects roadworthy).

When $T_{req}^{(k)}(\tau)>\tau$, then in the time interval for the posts of the analyzed subsystem the rendering of the required $k$ number of technical objects roadworthy is not possible.

Evaluation of operational availability of serviceability assurance subsystem SAS for carrying out the assigned task consists of determining the value of actual availability of the subsystem (for the required time interval of $\tau$ length as well as required $k$ number of technical objects which should be rendered roadworthy and/or stocked) and then comparing to the value of required availability the SAS should has in order for the assigned task to be carried out according to the relation

$$A_{OZ}^{(k)}(\tau)=A_O(\tau)\cdot Z^{(k)}(\tau)\geq A_{OZ_{req}}^{(k)}(\tau),\quad for\quad k\geq k_{req}(\tau) \tag{14}$$

If the value of operational availability of the SAS posts for carrying out the assigned task is lower than required, the assigned task at SAS posts may not be carried out properly (in the time interval of $\tau$ length it is not possible to render the required number $k_{req}(\tau)$ of technical objects roadworthy). The presented method makes it possible to analogically determine and evaluate the availability for carrying out the assigned service and repair task for both a serviceability assurance subsystem, group of posts or an individual post.

### 5.1 Determining the required availability of posts of the serviceability assurance subsystem in carrying out the assigned task

In the case of the serviceability assurance subsystem comprise $i = 1, 2, ..., p$ of groups containing $j = 1, 2, ..., q_i$ of posts linked to a given structure, it is possible to determine the required availability of posts at individual groups. Assuming that the $i$-th group comprises uniform posts are liked by threshold structure, the required availability of posts is assigned

$$A_{ij}=A_{ij_{req}}\Leftrightarrow A_{O_i}(\tau)=\sum_{c=k_i}^{q_i}\binom{q_i}{c}\cdot\left[A_{ij}\right]^{c}\cdot\left[1-A_{ij}\right]^{q_i-c}\geq A_{O_{i_{req}}}(\tau) \tag{15}$$

Figures 1 and 2 present values of required availability of posts at $i$-th group of serviceability assurance subsystem, determined in relation with number of posts of the $i$-th group as well as the value of parameters determining the assigned service and repair task (number of objects as well as the time of rendering them roadworthy).

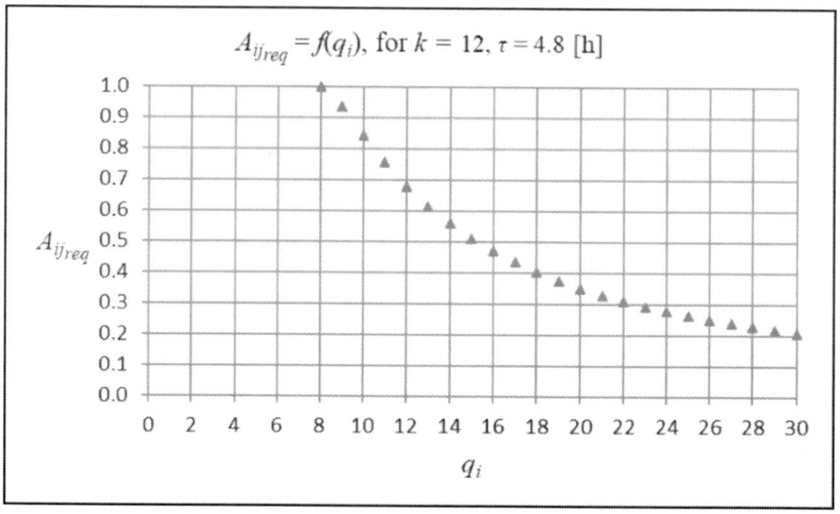


**Fig. 1** Required availability of individual post of functions of the number posts of the *i*-th group.

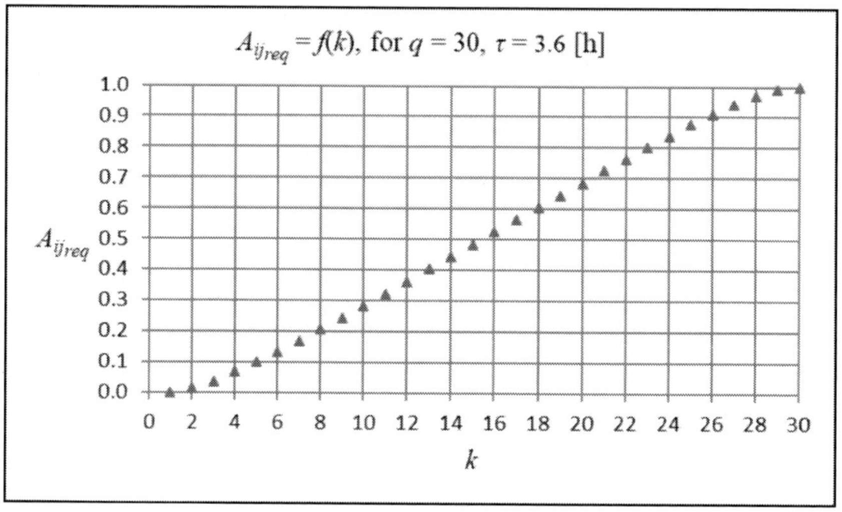


**Fig. 2** Required availability of individual post of functions of the number of objects rendered roadworthy.

## 5.2 Determining the required number of posts of the serviceability assurance subsystem in carrying out the assigned task

In the case of the serviceability assurance subsystem comprise $i$ = 1, 2, ..., $p$ of groups containing $j$ = 1, 2, ..., $q_i$ of posts linked to a given structure, it is possible to determine the required number of posts at individual groups. Assuming that the $i$-th group comprises uniform posts are liked by threshold structure, the required number of posts is assigned

$$q_i = q_{i_{req}} \Leftrightarrow A_{O_i}(\tau) = \sum_{c=k_i}^{q_i} \binom{q_i}{c} \cdot [A_{ij}]^c \cdot [1 - A_{ij}]^{q_i - c} \geq A_{O_{i_{req}}}(\tau) \tag{16}$$

Figures 3 and 4 present values of required number of posts at $i$-th group of serviceability assurance subsystem, determined in relation with availability value $G_{ij}$ of individual post of the $i$-th group as well as the value of parameters determining the assigned service and repair task (number of objects as well as the time of rendering them roadworthy).

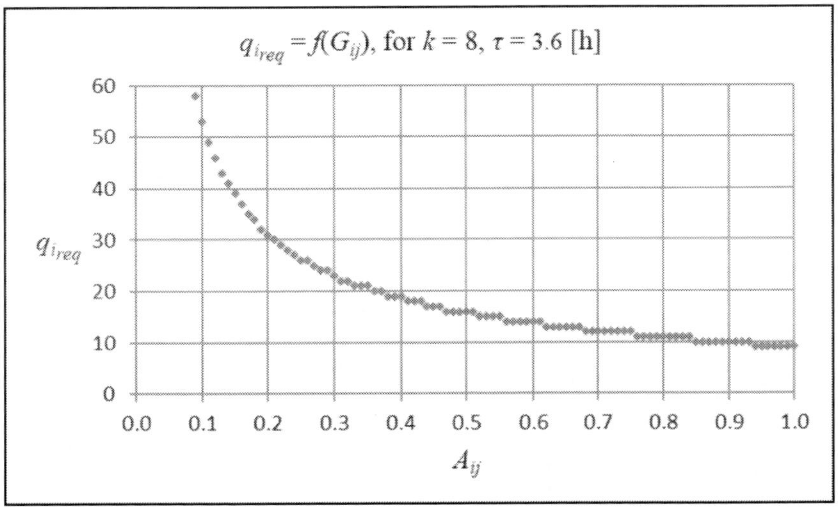


**Fig. 3** Required number of posts of the $i$-th group in the functions of availability of individual posts.

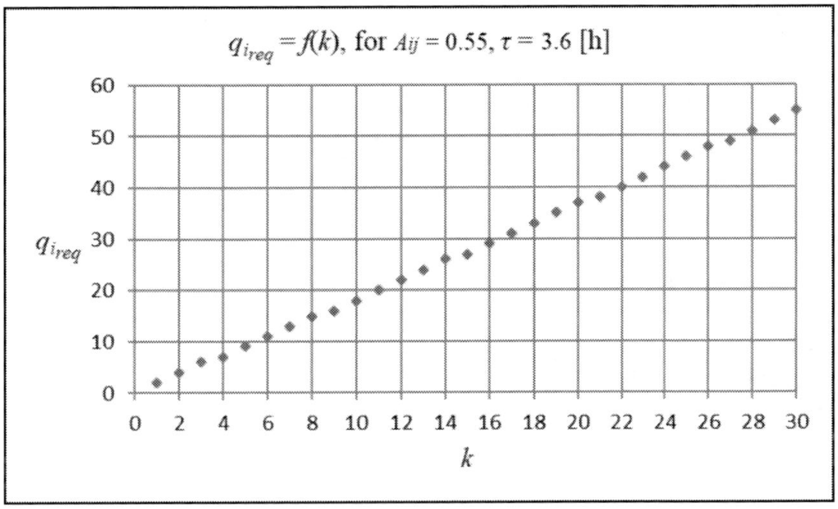


**Fig. 4** Required number of posts of the $i$-th group in the functions of number of objects rendered roadworthy.

## 6 Summary

On the basis of the method presented in the paper, it is possible to select (determine) the minimum required number and availability of posts of serviceability assurance subsystem so that the assigned service and repair tasks are carried out properly. It is completed on the basis of the selected criteria for evaluation when actual availability of serviceability assurance subsystem equals at least the required availability, e.g. for the number of objects rendered roadworthy $k = 8$, time of rendering roadworthy $\tau = 3.6$ [h] and individual post availability $A_{ij} = 0.8$, required number of posts of $i$-th group amounts to $q_{i_{req}} = 11$, then $A_{O_i}(\tau) = 0.795 \geq A_{O_{i_{req}}}(\tau) = 0.762$ (Figure 3).

The presented method may be utilized in order to evaluate operational availability of an individual post as well as a group of posts of a given type coupled with an appropriate

structure or a serviceability assurance subsystem consisting of various types of posts. Providing the required availability of serviceability assurance subsystem is possible due to:
– adjusting the number and structure of service and repair posts,
– adjusting the equipment at posts in order to, when necessary, facilitate carrying out of particular tasks at posts in various groups,
– implementation of posts (devices and tools) of higher reliability, durability, and productivity.

## References

1. D-Y. Chen, K. S. Trivedi, *Closed-form analytical results for condition-based maintenance* (Reliability Engineering and System Safety, Elsevier), (2002)
2. D-Y. Chen, K. S. Trivedi, *Optimization for condition-based maintenance with semi-Markov decision process* (Reliability Engineering and System Safety, Elsevier), (2005)
3. A. Gołąbek, *Metoda określania struktury systemu obsługi pojazdu samochodowego* (Zagadnienia Eksploatacji Maszyn), (1984)
4. L. Knopik, K. Migawa, *Optimal age-replacement policy for non-repairable technical objects with warranty* (Eksploatacja i Niezawodność – Maintenance and Reliability), (2017)
5. M. Marseguerra, E. Zio, *Optimizing maintenance and repair policies via a combination of genetic algorithms and Monte Carlo simulation* (Reliability Engineering and System Safety, Elsevier), (2000)
6. K. Migawa, *The evaluation of availability of the serviceability assurance subsystem posts* (Journal of KONBiN), (2012)
7. K. Migawa, L. Knopik, S. Wawrzyniak, *Application of genetic algorithm to control the availability of technical systems* (Engineering Mechanics, Institute of Thermomechanics Academy of Sciences of the Czech Republic), (2016)
8. J. Okulewicz, Z. Smalko, *Wyznaczanie technicznie uzasadnionej liczby stanowisk obsługowo-naprawczych* (Instytut Transportu Politechniki Warszawskiej), (1988)
9. M. Szubartowski, *Efficiency of repairs performed within the emergency repair subsystem of municipal bus transportation system* (Journal of KONES Powertrain and Transport), (2012)
10. M. Woropay, K. Migawa, P. Bojar, *The model of defining the efficiency of logistics subsystem posts in the transport system* (Journal of KONES Powertrain and Transport), (2010)
11. M. Woropay, J. Żurek, K. Migawa, *Model oceny i kształtowania gotowości operacyjnej podsystemu utrzymania ruchu w systemie transportowym* (Instytut Technologii Eksploatacji, Radom), (2003)
12. JK. Vaurio, *On time-dependent availability and maintenance optimization of stanby units under various maintenance policies* (Reliability Engineering and System Safety, Elsevier), (1997)

# Diagnostics – Experimental Analysis of Friction Pairs at Stick – Slip Sliding

*Aleksandr* Dykha[1,*], *Sergiy* Matyukh[1], *Tomasz* Kałaczyński[2]

[1]Khmelnytsky National University, 29016, 11, Instytuts'ka str., Khmelnitsky, Ukraine
[2]University of Science and Technology, Al. prof. S. Kaliskiego 7, 85-796 Bydgoszcz, Poland

**Abstract.** The movement of parts of friction pairs with small quantities and displacements is the cause of the so-called stick-slip movement. This is due to the instability of the process of formation and destruction of adhesive contact between the surfaces. In this case, the lubricant has a decisive influence on the processes of friction in the contact. In this work, based on the diagnostic and experimental complex, the dependences of the magnitude of the friction force on displacements for various lubricants were obtained. Statistical characteristics of the scattering parameters of friction force are obtained to estimate the vibrations of a moving body. It is received that the dispersion of the friction force relative to the average value decreases with increasing load and with dry and lubricated contact. On the basis of the obtained experimental results, it is possible to establish some quantitative characteristics of such a movement: the number of oscillations (jumps) per unit of time; the average movement in a contact before a shift in a contact; the average resistance force in a contact before a shift. The correct selection of lubricant and loads ensures minimum values of vibrations of the friction units of machines. This consequently improves the positioning accuracy of the machines.

## 1 Introduction

With small movements in speed (10 ... 100 mm/min), movement with jumps is observed in translational friction pairs (stick-slip movement). This is typical for sliding friction in machine tools and instruments, sliders, presses and other machines. On the type of movement in this case, the magnitude of the friction force and its changes have a decisive influence. At low speeds, there is a transition from the forces of friction of a fixed body to the forces of friction of motion and vice versa. This is due to the instability of the process of formation and destruction of adhesive contact between the surfaces. As a result, the positioning accuracy of the machine components decreases, additional dynamic loads are created, failures occur and the wear rate of the machine parts increases. In this regard, the study of the processes of uniformity of movement and accuracy of the installation movements is given great importance. In this paper, we studied the characteristics of the frictional interaction of translational friction pairs with various types of lubrication of the contacting surfaces.

## 2 Literature review

The influence of friction mode on the nature of the motion of mechanical objects is studied in a number of articles. The paper [1] presents a demonstration of an oscillator with sliding friction that exhibits very good agreement with a linear fall off in amplitude. The demonstration also confirms that sliding friction is proportional to the magnitude of the normal force. The paper [2] introduces a more detailed friction model, one that explicitly

* Corresponding author: tribosenator@gmail.com

considers deformation of and adhesion between surface asperities. Using probabilistic surface models for two nominally flat surfaces, the stick-slip model sums adhesive and deforming forces over all asperities. However, these studies do not take into account the damping properties of the lubricant. The lubricant has a decisive influence on the processes of friction and oscillations. In studies [3-4], methods for determining the parameters of sliding friction under damped oscillations are considered. Mathematical models and methods for identifying their parameters on the basis of experiments are proposed. The method of laboratory testing for friction has a great influence on the accuracy of the results of studying vibrations under friction. Works [5-6] propose various hardware (acoustic, electrical) to control the characteristics of sliding friction. These studies have a positive contribution to the study of the processes of friction, but they do not always determine the sufficient accuracy of the friction parameters. The result of friction on the law of motion is the accuracy of moving and positioning machine parts. In [7] the numerical simulations, the displacement of the moving part and the driving force of the controller are considered in cases of no friction, friction without stiction effect (sliding friction), and stick-slip friction. From these conditions, the influence of the friction on the sliding stability of the controller and the motion accuracy of the machine tool are analyzed and predicted. The study [8] gives an analysis of the basic friction models with tests and compares friction models in the literature: Coulomb friction model, Stribeck friction model, Dahl model, LuGre model, and the elastoplastic friction model on the same testbed . The experiments and models were done and the reasons for the difference in the basic of these models were investigated. Friction cannot be considered separately from modern surface engineering methods. The processes of friction depend on the properties of materials and coatings of the contacting bodies. In work [9] the influence of ionic nitriding on the processes of friction and wear under conditions of translational and swinging movement is considered. Resistance to wear of materials is also largely determined by friction processes. In work [10-12], the parameters of the friction process in the complex are presented in the model of wear proposed by the authors. The importance of laboratory and theoretical studies of friction parameters for practical applications follows from the literature review. For modern machines, friction is a necessary element that determines the accuracy of their operation. Thus, the study of friction processes requires more careful and detailed study.

## 3 Experimental equipment

For an experimental study of the friction force during the movement of a loaded plate on a lubricated and dry surface, the "LRX-E" model "Lloyd instruments" universal machine was used. The main characteristics of the testing machine are as follows:

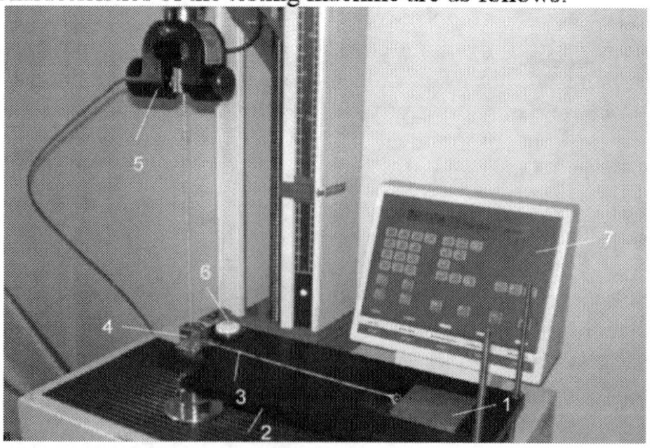


**Fig. 1.** Testing machine LRX–E.

1. Maximum force: 2.5 kN.
2. Error measuring force: 0.005% of maximum value.
3. The limits of change of the speed of movement: 0.1 ... 1020 mm / min.
4. The amount of movement: 1 ... 1500 mm.
5. Error of measurement of movement <5.

The testing machine (fig. 1.) Is a laboratory computerized complex allowing to carry out various tests in the "force-displacement" format. In our case, it was configured to test for sliding friction in accordance with the international standard ASTM D1894. When testing, a steel, ground (Ra = 1 µm) plate 1 (Fig. 1.) measuring 63.5 × 63.5 $mm^2$ and weighing 2N slid along the burnished surface of the table 2. The horizontal position of the table was controlled by level 6. The plate movement was carried out by means of tensile flexible filament 3 through block 4. The upper end of the thread was fixed in the grip 5. The grip 5 was fastened in the traverse and with the help of a screw was moved along the guide column. To measure the forces and displacements, a special strain gauge type LASERSCAN LR 01/2932 was used. The control of the testing process was carried out using the control panel 7 on the basis of a liquid crystal display. The software "Nexygen V4.5" and "Ondio V 4.5" was used to automate the test process with the specified parameters, automatically display the test results and statistical data processing. To study the friction force during the movement of a loaded plate on a flat surface, the following conditions were taken:

1. The speed of movement of the movable plate, mm/ min.
2. Distance of movement, mm.
3. Types of lubrication: 1-dry; 2- motor oil Formula Q8 (API- SJ/CD, SAE - 15W/40); 3-consistent oil Agip F1 CR MU3 (Castrol).
4. Loading mode: 1 – 2N; 2 - 7.5 N; 3 - 13.5 N.

## 4 Results and Discussion

The test results are presented in the graphs on fig. 2-4. Depending on the type of disturbing forces, the fluctuations of the friction force can be either deterministic (often periodic) or random. In our case, the reasons for the fluctuation of the friction force are the following: the random type of interaction of the roughness of the contacting surfaces (as a result, there is a random process of formation and destruction of frictional bonds between the surfaces); time dependence and different intensity of molecular adhesion forces (adhesion) between dry and lubricated surfaces; adhesive properties of lubricants; non-uniform distribution of the lubricant layer over the sample surface thickness. As a result, there is a stick–slip movement of a moving specimen (at relatively low sliding speeds) with random transitions from static friction to movement friction.

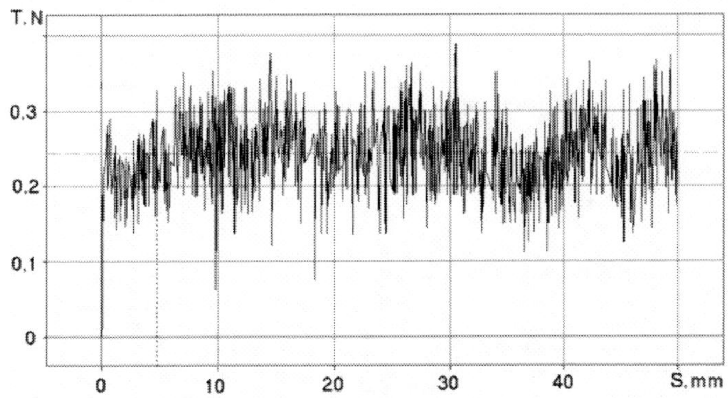


$Q = 2N \; (\bar{f} = 0{,}12 \, ; \bar{T} = 0{,}24N)$

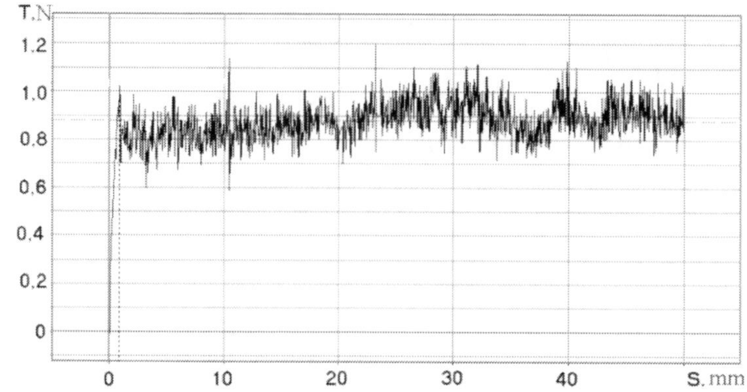


$Q = 7{,}5N \; (\bar{f} = 0{,}117; \bar{T} = 0{,}88N)$

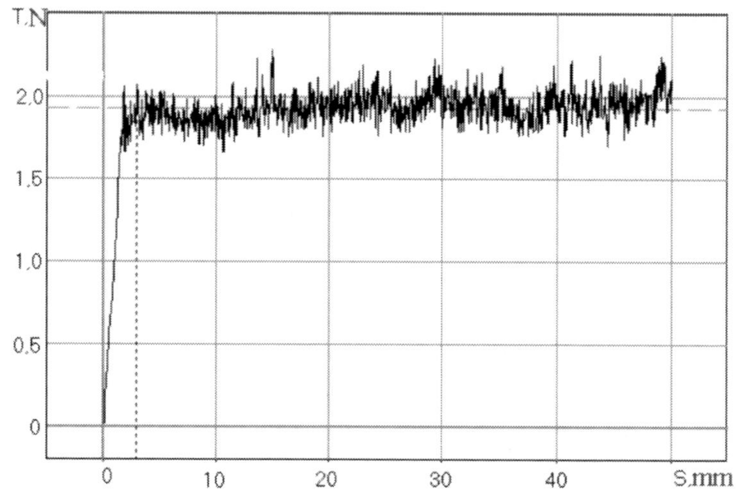


$Q = 13{,}5N \; (\bar{f} = 0{,}14; \bar{T} = 1{,}9N)$

**Fig. 2.** Dependence of the friction force T on the displacement S during dry contact.

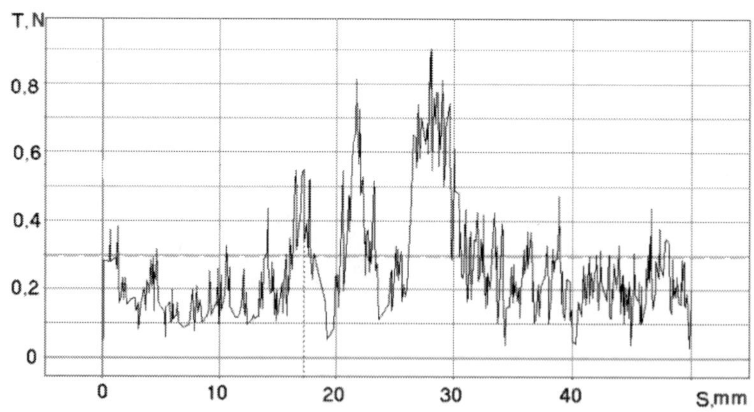


$Q = 2N \; (\bar{f} = 0{,}147; \bar{T} = 0{,}295N)$

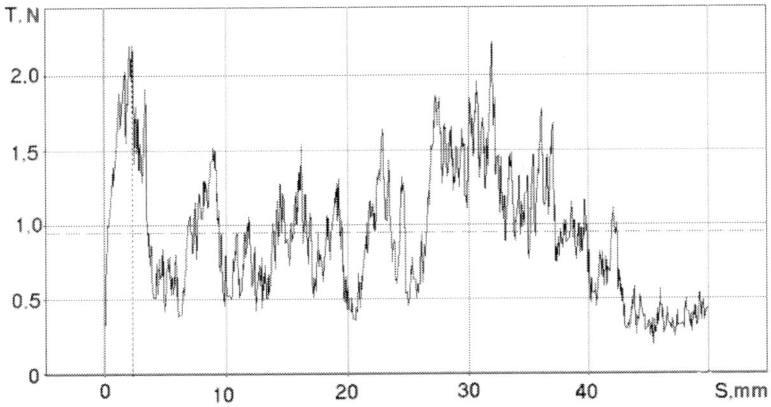


$Q = 7{,}5N \ (\bar{f} = 0{,}12 \,; \overline{T} = 0{,}9N\,)$

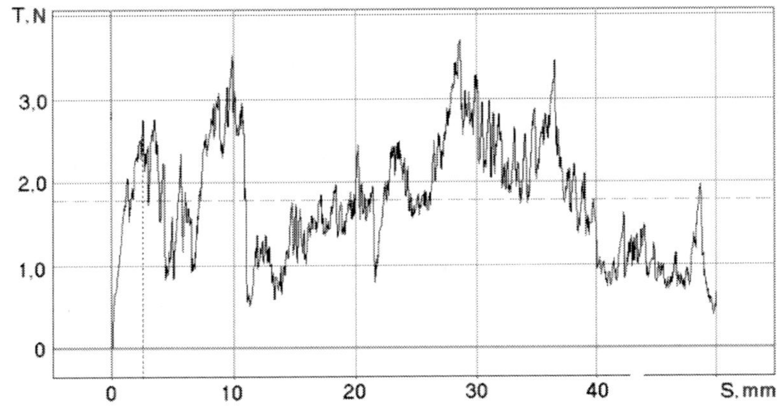


$Q = 13{,}5N \ (\bar{f} = 0{,}14 \,; \overline{T} = 1{,}9N\,)$

**Fig. 3.** Dependence of the friction force T on the displacement S with motor oil Formula Q8.

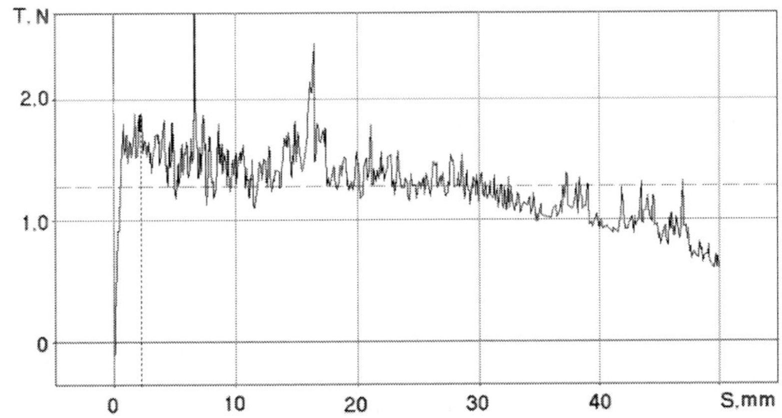


$Q = 2N \ (\bar{f} = 0{,}64 \,; \overline{T} = 1{,}28N\,)$

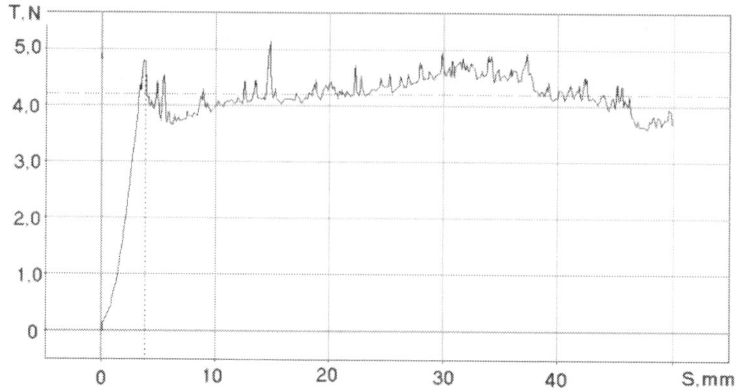


$Q = 7{,}5N$ ($\bar{f} = 0{,}56$ ; $\bar{T} = 4{,}22N$)

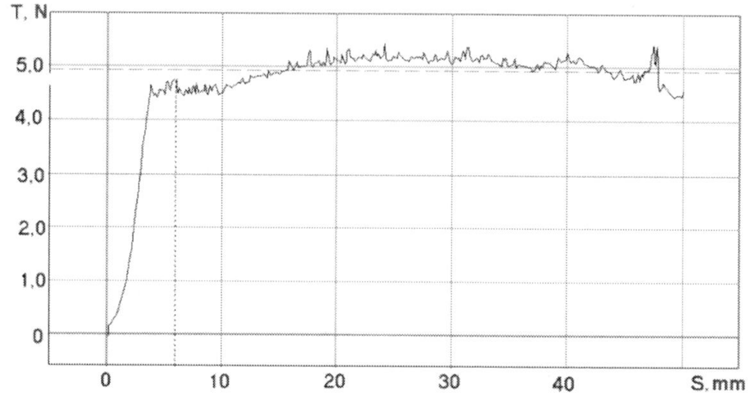


$Q = 13{,}5N$ ($\bar{f} = 0{,}36$ ; $\bar{T} = 4{,}92N$)

**Fig. 4.** Dependence of the friction force T on the displacement S with grease.

The friction force can take any value in a certain range. Therefore, the random process of friction force change should be characterized not by amplitude, frequency and phase, but by the mean value $\bar{T}$, standard deviation $\sigma_T$, coefficient of variation $\upsilon_T$, minimum $T_{min}$ and maximum values $T_{max}$ and the range of change of the friction force $\Delta T$.

To determine these characteristics, a statistical processing of the test results obtained on a testing machine was carried out. The results of statistical processing are shown in table 1.

**Table 1.** Statistical parameters of friction force changes.

| Lubricant | $Q$, N | $T_{min}$, N | $T_{max}$, N | $\Delta T$, N | $\bar{T}$, N | $\sigma_T$, N | $\upsilon_T$ |
|---|---|---|---|---|---|---|---|
| Dry | 2 | 0,08 | 0,4 | 0,32 | 0,24 | 0,078 | 0,325 |
| | 7,5 | 0,59 | 1,3 | 0,71 | 0,88 | 0,117 | 0,132 |
| | 13,5 | 1,63 | 2,27 | 0,64 | 1,9 | 0,14 | 0,073 |
| Motor oil Formula Q8 (API- SJ/CD, SAE - 15W/40) | 2 | 0,04 | 0,89 | 0,85 | 0,295 | 0,179 | 0,607 |
| | 7,5 | 0,26 | 2,25 | 1,98 | 0,9 | 0,508 | 0,564 |
| | 13,5 | 0,41 | 3,81 | 3,4 | 1,9 | 0,72 | 0,378 |
| Agip F1 CR MU3 (Castrol) | 2 | 0,6 | 2,4 | 1,8 | 1,28 | 0,293 | 0,229 |
| | 7,5 | 3,64 | 5,13 | 1,49 | 4,22 | 0,297 | 0,070 |
| | 13,5 | 4,42 | 5,49 | 1,07 | 4,92 | 0,27 | 0,055 |

An analysis of the results obtained by studying the friction force indicates the following. With a dry contact of the ground plate with the surface of the table, relatively low friction forces take place compared to even the lubricated contact. This can be explained by the low adhesion of the surface of the steel plate to the surface of the table, covered with a layer of protective oxide film.

When the plate moves along the motor oil layer, the friction force in comparison with the dry contact should decrease, but the results suggest that it is approximately at the same level as in the case of dry friction. Obviously, this is due to the additional resistance to movement due to the high adhesion of the lubricant to the surfaces at low sliding speeds. This effect is even more pronounced with the grease.

The dispersion of the friction force relative to the average value decreases with increasing load and with dry and lubricated contact. This is due to the increase hardness in contact with increasing load.

Stick-slip change in friction force is associated with the transition from friction of non-mobility to friction of motion. On the basis of the obtained experimental results, it is possible to establish some quantitative characteristics of such a movement: the number of oscillations (jumps) per unit of time $\nu_c$; the average movement in a contact before a shift in a contact $s_c$; the average resistance force in a contact before a shift $T_c$. As a result of the graph analysis of fig 2 - 4, the following numerical values of these characteristics are established (Table 2).

The analysis of the obtained results shows that the greatest frequency of friction force oscillations occurs during dry contact. In this case, weak adhesion forces act and changes in the friction force are repeated more often than for contact with a lubricant. When using a lubricant, a large force is required to shift the contact layer and a correspondingly longer time for shear is required. Plastic lubricants at the same time have the best damping properties, since jumps of friction during their use are performed with the least frequency. However, the coefficient of friction is quite large (up to 0.5) due to the high adhesive properties of greases.

**Table 2.** Characteristics of the hardness of the contact layer for moving loaded plate.

| Lubricant | $Q$, N | $\nu_c$, sec$^{-1}$ | $S_c$, μm | $T_c$, N |
|---|---|---|---|---|
| Dry | 2 | 5,3 | 156 | 0,15 |
| | 7,5 | 7,7 | 109 | 0,26 |
| | 13,5 | 7,4 | 114 | 0,29 |
| Motor oil Formula Q8 (API- SJ/CD, SAE - 15W/40) | 2 | 2,9 | 286 | 0,28 |
| | 7,5 | 4,1 | 204 | 0,84 |
| | 13,5 | 4 | 208 | 0,97 |
| Agip F1 CR MU3 (Castrol) | 2 | 2,9 | 285 | 0,36 |
| | 7,5 | 2 | 417 | 0,57 |
| | 13,5 | 2,1 | 400 | 0,45 |

The obtained results of the quantitative analysis of the stick-slip movement with various lubricants are recommended for practical use in industry. The correct selection of lubricant and loads ensures minimum values of vibrations of the friction units of machines. This consequently improves the positioning accuracy of the machines.

## 5 Conclusions

On the basis of the laboratory diagnostic complex "LRX-E", studies of the characteristics of friction during movement with a low speed of the loaded plate on the working surface in the presence of lubrication were carried out. As a result of processing the experimental data, statistical characteristics of the friction parameters were obtained, which characterize the effect on them of the load and lubrication mode. The quantitative characteristics of the plate motion, which characterize the deformation properties of the contact layer, are established.

## References

1. M. Kamela, The Physics Teacher, V. **45**, 110 (2007)
2. M. Bengisu, A. Akay, J. Acoust. Soc. Am. 105 (1), (1999)
3. A. Dykha, V. Aulin, O. Makovkin, S. Posonskiy, Eastern- European Journal of Enterprise Technologies, V. **3**, 7 (87), (2017)
4. A. Dykha, A. Kuz'menko, J. Frict. Wear, V. **37**, 4, (2016)
5. V. Aulin, A. Hrynkiv, A. Dykha, M. Chernovol, O. Lyashuk, S. Lysenko, Eastern-European Journal of Enterprise Technologies, V. **2**, 1(92), (2018)
6. A. Dykha, S. Matyukh, *Triboacoustic diagnostic fixed joints of machines*, MATEC Web Conf., 182, (2018)
7. Thi-Na Ta, Yunn-Lin Hwang, Jeng-Haur Horng, Jurnal Tribologi, 19, 107-120, (2018)
8. Y. F. Liu et al.: Experimental comparison of five friction models on the same test-bed, Mech. Sci., 6, 15–28, (2015)
9. P. Kaplun, O. Dykha, V. Gonchar, Materials Science, V. **53**, 4 (2018)
10. A. Dykha, R. Sorokatyi, O. Makovkin, O. Babak, Eastern-European Journal of Enterprise Technologies, V. **5**, 1 (89), (2017)
11. M. Żółtowski,M. Liss, B. Żółtowski, J. Melcer, *Truss harbor cranes modal design elements research,* Czasopismo: Polish Maritime Research, **22, 4(88)**, (2015.)
12. J. Wilczarska, E. Kuliś, M. Łukasiewicz, Ł. Fornal, N. Dluhunovych, *The assessment of the impact of the chosen exploational conditions of hydraulic arrangement on the working liquid condition,* Matec WoC, **182**, 1-12, (2018)

# Lapping flat surfaces with forced dosing of abrasive suspension

*Maciej* Gniot[1,*], *Adam* Barylski[2]

[1]UTP University of Science and Technology, Faculty of Mechanical Engineering
[2]Gdańsk University of Technology, Faculty of Engineering

**Abstract.** The issues concerning the finishing of flat ceramic seals are presented in the paper. The treatment requirements demanded from the elements of that type cause, that necessary is application of grinding-in in the technological process. Such a precise abrasive treatment of flat surfaces of that type, requires the use of an appropriate suspended matter or abrasive paste. Many factors connected with metal grinding-in shield, components of abrasive suspended matter and the treatment's technological conditions, have an impact on the treatment's result. The manner and intensity of the suspended matter's proportioning in the grinding-in are, are also of decisive nature. A standard, free manner of proportioning, in a continuous manner most often dipping or dropping, results in big losses of abrasive material. A part of abrasive micro-grains is quickly removed from the operating grinding-in shield by the rotating rings leading separators with treated elements and does not take part at all in machine cutting. It considerably increases the tool costs and has an unfavourable impact on the natural environment. Implementation of the forced abrasive suspended matter's proportioning at the time of the machine grinding-in, considerably reduces these losses.

## 1 Innovative proportioning system

The developed innovative proportioning system makes it possible to control appropriately the volume and frequency of the abrasive suspended matter's applying on the specific place on the tool's operating surface. It also makes it possible to prepare the suspended matter of optimum viscosity. The constructed test stand for one-side grinding-in of flat surfaces, is based on the construction of the lapping machine Abralap 380 (Fig. 1).

**Fig.1**. Forced system of proportioning and applying the abrasive suspended matter: 1- system of constant mixing, 2- system of the suspended matter's application, 3- control system, 4- camera registering the process.

* Corresponding author: maciej.gniot@utp.edu.pl

The system of proportioning and applying of the abrasive suspended matter's applying, is divided into two systems. The first of them answers for preparation and supplying of the abrasive suspended matter (Fig. 2 and 3).

**Fig. 2**, System of the abrasive suspended matter's preparation: 1- compressor, 2- power transmission of the mixer's rotor, 3- pressure regulating device regulator inside the tank, 4- the tank's cover, 4- suspended matter's tank, 5- bowl of the tank.

The system of the abrasive suspended matter's applying (Fig. 4), makes it possible to apply a layer of the assumed thickness (expressed per one rotation of the lapping tool). The system also has the option of controlling the volume of the applied spot of abrasive suspended matter, its shape and placement of the shield (Fig. 5). The process of proportioning and applying, is controlled via the electronic system.

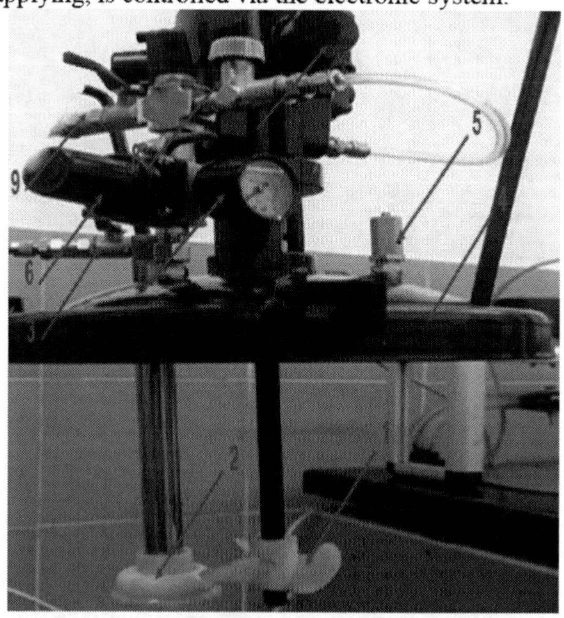

**Fig. 3**. System of continuous mixing of the abrasive suspended matter: 1- mixer's screw, 2- dragon, 3- feeding of the applying apparatus with the abrasive suspended matter, 4- tank's cover , 5- safety valve, 6- air pressure regulator, 7- pressure gauge, 8- mixer's rotational speed regulator, 9- compressed air terminal.

As it may be observed, the size and shape of the stain depend on the pre-set settings. It may also be found, that the system makes it possible to proportion the volume of a single suspended matter's dose, what translates into the possibility of planning the total consumption of the abrasive suspended matter at the moment of planning the elements' grinding-in process, taking into consideration their properties. Maintaining of a certain, assumed thickness of the applied layer on the lapping tool's shield during the course of the whole grinding-in process, is also possible [1- 4].

**Fig. 4.** System applying the abrasive suspended matter on the lapping tool: 1- shape of the stain made of the abrasive suspended matter, 2- direction of the lapping tool's rotation, 3- applied layer of the suspended matter.

**Fig. 5** Possible location of the applied abrasive suspended matter's layer .(1- system of the applied layer location's change on the lapping tool's shields, 2- possible stain's location on the shield, 3- possible stain's location on the shield of a specified size, 4- possible stain's location on the shield with hanging area).

## 2 Selection criteria of the research programme

Setting about to own studies, requires assessment of technical possibilities concerning making measurements of the selected elements, and ensuring an appropriate test stand. Identification, within the scope of the volumes of tests possible to be realized in order to conduct a correct process on their basis, is the necessary element of preparations for basic tests' conducting. Also, an estimate time of tests' realization is to be assessed and the financial costs connected with such an undertaking is to be determined. For that purpose, based on the criteria discussed in the previous section, that is efficiency and informativity of individual research programmes [5, 7, 11, 13], there shall be conducted the process of identification. In order to make comparison possible, the number of the tested factors and the necessary number of the tests is to be determined, just as to make it possible to obtain the sought, highly reliable information [6, 9, 14]. In Fig. 6, there is presented the diagram of proceeding at the time of the individual solutions' usability valuation, the selected research programmes.

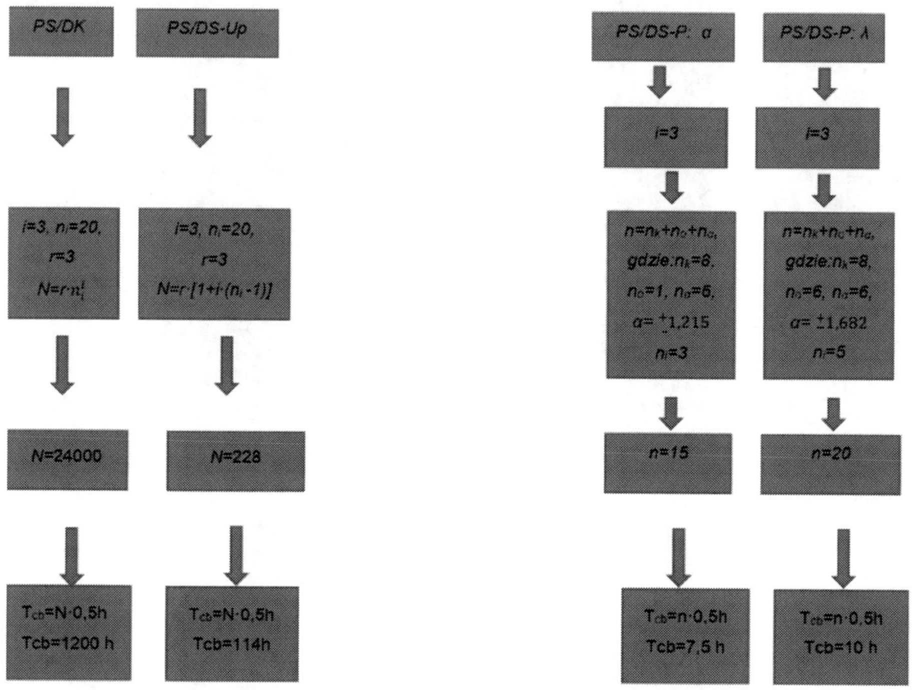


**Fig. 6**. Comparative diagram of research programmes.

The tested factors have been determined theoretically, based on the analysis of the applied abrasive suspended matter's proportioning manners.

Analysing the potential possibility of the PS/DK programme's use for own studies it was found, that the programme in its assumption uses all the possible configurations of tested factors (that is, each one with each one), what of course translates into the following possibilities. Assuming that i=3 tested factors, number of tests ni = 20, and r = 3 three repetitions, there was received the total number of tests that should be conducted N=24000. It is a very big number, assuming that the time of one single test t = 0,5 h, what gives us the realization time on the level of Tcb = 1200 h. Assessing the usability of the PS/DK programme it was found, that the programme's informativity strongly depends on the number of the tested factors and their arrangement in the intervals [$x_{max}$, $x_{min}$], what influences a long tests' time.

Also the PS/DS.-Up programme has been analysed from the point of its possible use for the tests. This programme uses other possibilities, namely it is based on selecting the tested factors X and determining their central values. The central values determined in such a manner do not change at the time of subsequent tests concerning the impact of a single X factor. In order to make it possible to determine the central value, for the first there should be conducted the preliminary analysis of the tested object. The PS/DS-Up programme has the character of the procedure unequivocally determined prior to the tests' commencement, following determination in the first place of the programme central values. Informativity of the PS/DS-Up programme is slightly limited as compared to the PS/DK programme, and to a considerable degree it depends on the arbitrary selection of the central values. The number of the required measurements in the PS/DS-Up programme amounts to N=228 with the same assumptions for i = 3, ni = 20, r = 3. The total time of the tests at that programme's assumptions, has been assessed for Tcb = 114 h.

Programmes PS/DS-P: α and the programme PS/DS-P:λ in their assumptions make use of the nonlinear model in the form of the second degree multinomial. The second-degree multinomial influences:

- limitation of the number of measurements,
- limits complications connected with statistical calculations and the recourse analysis.

The basic feature of the programmes PS/DS-P:α and PS/DS-P:λ is implementation of special compositions (systems) of the tested factors' values ensuring the so-called orthogonality notion which concerns the PS/DS-P:α and the notion of rotatability connected with the programme PS/DS-P:λ. During analysis usability of the above programmes for the tests it was found, that while choosing the PS/DS-P:α programme, the tests shall be realized at four levels, and the total number n of the systems for measurement's realization n = 15, the time of the tests' realization with the use of that programme shall amount to Tcb = 7,5 h. In the programme PS/DS-P:λ, the tests shall be realized at five levels, and the total number n of systems for the measurements' conducting n = 20, the time of the tests' realization with the use of that programme shall amount to Tcb = 10 h.

Finally, for conducting of the tests concerning grinding-in of flat ceramic elements with the forced proportioning of the abrasive suspended matter, there has been selected the programme PS/DS-P:λ as its time consumption amounts to Tcb = 10 h, informativity is relatively high, and the model described in this programme shall be the second degree multinomial, what shall allow for graphic interpretation of the function's form [5, 6, 9, 14].

## 3 Development of the experiment's results

The developed and constructed abrasive suspended matter's proportioning system, makes it possible to conduct experimental studies of grinding in of flat ceramic sealing elements, with the use of the static, determined, selective, multi-factor, second degree quasi-rotatable PS/DS-P: λ, of spheroidal information's distribution plan of the tests [15]. Such a solution ensures stability of the recourse function's assessment in a given environment of the plan's central point. The programme is characterised with relatively high informativity and efficiency, what from the point of view of the tests' costs consumption, is an essential factor. The experiment, in case of three variable tested factors, requires conducting only n = 20 tests [7, 8, 10, 11, 12].

As the variable factors, there have been selected the: percentage content of abrasive grains (weight share) K[%] = $x_2$, dose of abrasive suspended matter applied on the shield for 20 minutes Vs[ml/20 min] = $x_3$ , and viscosity of the mixture of the suspended matter's liquid components Le[mPa·s] = $x_4$. The variability range of the tested factors has been determined based on the literature and preliminary studies. For the tests, as the abrasive grain, there has been used the silicon carbide 98C F400/17. The samples made of $Al_2O_3$ ceramics should be subject to grinding in. The values of the output factors Y in the experiments, are the variables of random character. Such a premise makes it possible to assume, that the occurring interferences U and the constant conditions C, are the parameters seemingly fixed for determined tests' conditions. The function of the object has been determined in the multi-dimensional factor space, as the recourse equation: [5, 15, 16].

$$Y = F(x_1, x_2, x_3, \dots, x_k) \tag{1}$$

where:
k – is the number of the analysed variables.

Determined coefficients B = [b1, b2,…, $b_k$] concern the recourse equation:

$$Y = b_1x_1 + b_2x_2 + b_3x_3 + b_4x_4 + b_5x_2^2 + b_6x_3^2 + b_7x_4^2 + b_8(x_2 * x_3) + b_9(x_2 * x_4) + b_{10}(x_3 + x_4) \tag{2}$$

where:
value of the output factor from the model, while the independent variables: $x_1 = 1$, $x_2 = K$, $x_3 = Vs$, $x_4 = Le$.

The results of the conducted tests are to be subject to statistical analysis. Determining of the quadratic recourse equations describing the mass decrement Ubm and linear Ubl of the treated ceramic elements in the function of variable input data: K, Vs and Le, is the basic task. Estimation of coefficients b = [$b_1$, $b_2$, …, $b_k$] of the recourse equation is the first step. If by *X* we shall mark the experiment plan's matrix, and by *Y* the column of the measurements results, then vector *B* estimates with the use of the matrix formula in the form of

$$B^T = (X^TX)^{-1}X^TY,$$

(3)
where:
$X^T$ means the symbol of transaction's matrix *X*, and $X^{-1}$ is the matrix reverse to the matrix *X*. Estimation of vector B is based on the least squares' method. The unbiased estimator of variance $\delta^2$ of the random component, is the statistics $s^2$ determined with the formula:

$$s^2 = \frac{1}{n-k-1}\sum_{i=1}^{n}\left(y(i) - yt(i)\right)^2 \tag{4}$$

where:
$y(i)$ is the $i$-th measurement point (i-th point of the experiment's plan), $yt(i)$ is the value of the recourse function in the i-th point, n is the number of the plan's points, while k+1 is the number of the recourse equations' coefficients. Formula (4) in the matrix form is the following:

$$s^2 = (Y^TY - b^TX^TY)/(n - k - 1) \tag{5}$$

The assumption on normality of the random component's distribution, makes it possible to determine the confidence intervals and making use of the appropriate statistical tests for coefficients bi (i = 1, 2, … , k) of the recourse equation. The multiple correlation coefficient *R* is the parameter describing the quality of approximation with the empirical dependence equation P. The coefficient *R* measures the dependence (correlation) level of the variable *Y* on the independent variables $X_1$, $X_2$, … , $X_n$. The square of the correlation coefficient is the coefficient of determination $R^2$, where $yt(i)$ is the value of the function in the i-th point of the experiment's plan, $ye(i)$ is the measured (empirical value) value, while $ys$ is the mean value calculated from all the experiment's results. So:

$$R^2 \sum_{i=1}^{n} \frac{(ye(i)-yt(i))^2}{\sum_{i=1}^{n}(ys-yt(i))^2} \quad (6)$$

The closer is the value of the correlation coefficient $R$ , the higher is the degree of the variable explained by the recourse function. The value of $R$ close to zero, means the lack of dependence between the variable $Y$ and the set of variables $X_1, X_2, \ldots, X_n$. In order to check, whether the correlation coefficient is significantly different than zero, the test of recourse's significance has been conducted. In this case, the zero test's hypothesis, has the form of:

$$H_0: R = 0 \quad (7)$$

The testing statistics for that hypothesis, is expressed with the formula:

$$F = \frac{n-k-1}{k} \frac{R^2}{1-R^2} \quad (8)$$

Random variable F has the F-Snedecor distribution, of the degrees of freedom $d_{f1}=k$ (numerator) and $d_{f2} = n-k-1$ (denominator). The critical value $F_{\alpha,df1,df2}$ for the pre-defined level of significance α is to be read from the statistical tables. At the time of the multi-factor recourse equation's construction, the tested variables which may have an impact on the value of the dependent variable Y, should be taken into consideration. In order to check, which ones from the variables $X$ influence significantly the variable $Y$ , it shall be necessary to conduct the test of significance for all the recourse coefficients of the defined equation. The zero hypothesis is then in the form:

$$H_0: b_i = \beta_i \quad (9)$$

The testing statistics is:

$$t = \frac{b_i-\beta_i}{S_{b_i}} \quad (10)$$

where:
$b_i$- i-th recourse equation's coefficient (2),

$$S_{b_i} = s\sqrt{d_{ii}}. \quad (11)$$

where, s is the standard deviation's equation (2), while $d_{ii}$ is the i-th diagonal value of the matrix $(X^TX)^{-1}$.

The statistics t determined with the formula (10), while assuming genuineness of the $H_0$ hypothesis, has the distribution of t-Student of n-k-1freedom degrees. The critical area for that hypothesis is: $(-t_{n-k-1,\alpha}, t_{n-k,\alpha})$. Such value $t_{n-k-1,\alpha}$ is determined from the table t-Student for n-k-1 levels of freedom and the level of significance α.
Dispersion of the value of coefficient $b_i$ (i=1, 2, …, k) is described by the confidence interval:

$$\left(b_i - t_{\propto s_{b_i}}, b_i + t_{\propto s_{b_i}}\right), \text{ dla } i=1, 2, \ldots, k \quad (12)$$

where:

$$P\left\{b_i - t_{\propto s_{b_i}}, b_i + t_{\propto s_{b_i}}\right\} = 1-\propto \quad (13)$$

while 1-α is the level of confidence.

## Summary

The manner of applying the abrasive suspension matter has an essential impact on its consumption during the process's realization. It influences the treatment's effectiveness, the total cost of the process' realization, and the costs connected with utilization of the products' remaining following the completed process. Constructed forced proportioning and applying abrasive suspended matter's system together with the control system, makes it possible to implement limitations connected with the abrasive substance's consumption. In order to confirm that assumption, the research experiment is to be conducted obtaining the empirical results. The experiment should be conducted with the use of the research programme of the PS/DS-P:λ programme. The obtained results are to be put through statistical analysis, and the object's function in the multi-dimensional factor space determined as the recourse equation. The obtained equations shall make it possible to create area diagrams of the impact of the selected factors, that is: K, Vs, Le on realization of the process of the elements' grinding-in made of $Al_2O_3$ ceramics.

## References

1. A. Barylski, *Docieranie powierzchni płaskich na docierarkach* (Gdańsk, Wydawnictwo Politechniki Gdańskiej, 2013)
2. A. Barylski, M. Gniot, *Wpływ zawiesiny ściernej dawkowanej w sposób wymuszony na wydajność docierania jednotarczowego elementów ceramicznych* (Mechanik, 2018)
3. M. Gniot, A. Barylski, K. Migawa, *System dawkowania zawiesiny ściernej w docieraniu powierzchni płaskich* (Mechanik, 2017)
4. M. Gniot, A. Barylski, *Hydrodynamiczne dawkowanie zawiesiny ściernej w docieraniu jednotarczowym powierzchni płaskich* (Mechanik, 2016)
5. Z. Polański, *Metody optymalizacji w technologii maszyn* (Warszawa, PWN, 1977)
6. *Novye idei vplanirowanii eksperimenta,* red. V. V. Nalimov, (Moskva, Nauka, 1969)
7. V. V. Nalimov, *Teoria eksperymentu* (Moskva, Nauka, 1971)
8. V. V. Nalimov, N. A. Czernowa, *Statystyczne metody planowania doświadczeń eksperymentalnych* (Warszawa, WNT, 1967)
9. J. Kiefer, *Ann. Math.* (Statistics, 1961)
10. H. Ahrens, *Analiza wariancji* (Warszawa, PWN, 1970)
11. B. Kacprzyński, *Planowanie eksperymentów* (Warszawa, WNT, 1974)
12. G. E. P. Box, K. B. Wilson, *J. Royal Stat. Soc.* (Ser. B, 13, 1951).
13. V. V. Federov, *Teoria optymalnego eksperimenta.* (Maskva, Nauka, 1971)
14. J. Kiefer, *Proc.4-th Berkley Symp.* (Match. Statistics and Prod. 1, 1961)
15. G. Box, J. Hunter, *Multifactor experimental designs for exploring response surfaces* (Ann. Match. Statist., 1957)
16. K. Mańczak, *Technika planowania eksperymentu* (Warszawa, WNT, 1976)

# Research of characteristics of environmental energy carriers by simulation of operating load in laboratory conditions

*Romana* Janouskova[1,*], *Jerzy* Kaszkowiak[2], *Marek* Halenar[3], *Martin* Zachar[4], and *Jozef* Nosian[1]

[1]Slovak University of Agriculture in Nitra, Faculty of Engineering, Department of Transport and Handling, Tr. A. Hlinku 2, 949 76 Nitra, Slovak Republic
[2]UTP University of Science and Technology in Bydgoszcz, Al. prof. S. Kaliskiego 7, 85-796 Bydgoszcz, Poland
[3]ARRIVA Slovakia a.s., Sturova 72, 949 01 Nitra, Slovak Republic
[4]University of Defence Brno, Faculty of Military Technology, Department of Combat and Special Vehicles, Kounicova 65, 662 10 Brno, Czech Republic

**Abstract.** The focus of this sceintific article is a laboratory research on the qualities of environmental energy carriers by simulating the operational load, focusing on changes of physico-chemical qualities of ecological hydraulic and transmission fluid used in agricultural tractors. In laboratory setting by using a simulator, it was possible to simulate different operational load and also track changes in qualities of hydraulic fluid by changing the flow in precise intervals. The hydraulic-transmission fluid used in the laboratory setting was the MOL Farm NH Ultra.

## 1 Introduction

Ecological fluids used in mobile energy products operating in an environmentally sensitive environment are subject to high requirements that manufacturers have to take into account during their development in order to improve their quality while reducing the burden on the environment [1, 2, 3]. This is a natural consequence of efforts to increase the efficiency of production processes primarily due to a reduction in production cost, reduction in energy consumption with emphasis on reducing environmental impacts caused by production it-self, but also other dependent aspects [4]. A comprehensive analysis of the effect of biological fluids and mixtures thereof has already been dealt with author [5]. The change in the physico-chemical properties of the transmission-hydraulic fluid is an important indicator of the technical condition of the hydraulic pump and has a fundamental influence on the operation of individual elements of the hydraulic circuit. The oil level is evaluated according to either pre-scheduled intervals or continuously [6, 7, 8]. Proper technical diagnosis is provided high security and reliability with a view to extended maintenance cycles and limiting consequential damages. Objective technical status should be determined without dismantling and disruption [9]. It is appropriate to simulate the operational load in laboratory conditions for accelerated tests of ecological energy carriers. This also optimizes the laboratory test time and the possibility of multiple repetitions without the influence of the external environment which cannot be achieved in operational conditions [10, 11, 2]. After the end of operational load of ecological liquids, we analyze their physico-chemical properties and influence of used working fluid on the change of flow properties of the hydraulic pump.

*Corresponding author: xjanouskova@uniag.sk

## 2 Material and methods

A laboratory test facility which allows laboratory tests to be performed based on signals recorded during field tests was designed in order to simulate the operational load of the gear pump used in the tractor's hydraulic system when applying the tested ecological-hydraulic fluid [3]. Experimental set of pressure measurement with numerical simulation in engines under laboratory conditions shortens the time required for realization of operational measurements [12]. The basic prerequisite for proper functioning and effective care of hydraulic fluids is a suitably chosen methodology for testing fluids with monitoring of the level of contamination of working fluid [7, 13, 1]. The research was focused on testing the ecological hydraulic transmission fluid after working 500 hours on the basis of standardized conditions according to the Vickers test.

**Table 1.** Basic technical data of universal biodegradable fluid.

| Parameter | Unit | Value |
|---|---|---|
| Kinematic viscosity at 40 °C | $mm^2/s$ | 67,5 |
| Kinematic viscosity at 100 °C | $mm^2/s$ | 18,2 |
| Density at 15 °C | kg/m | 931 |
| Flash point | °C | 212 |
| Pour point | °C | -48 |

**Table 2.** Parameters of hydraulic pump OHD 17 (Katalóg zubových čerpadiel, 2013).

| Parameters | | Label | Unit | Value |
|---|---|---|---|---|
| Actual geometric volume | | $V_g$ | $dm^3$ | $17,24\times10^{-3}$ |
| Rotations | nominal | $n_n$ | 1/min | 1500 |
| | minimum | $n_{min}$ | 1/min | 350 |
| | maximum | $n_{max}$ | 1/min | 3200 |
| Inlet pressure | minimum | $p_{1min}$ | MPa | -0,03 |
| | maximum | $p_{1max}$ | MPa | 0,05 |
| Outlet pressure | max. permanent | $p_{2n}$ | MPa | 29 |
| | maximum | $p_{2max}$ | MPa | 31 |
| | peak | $p_3$ | MPa | 32 |

| | | | |
|---|---|---|---|
| Nominal input flow (min.) at $n_n$ and $p_{2n}$ | $Q_n$ | $dm^3/min$ | 23,2 |
| Maximum flow at $n_{max}$ and $p_{2max}$ | $Q_{max}$ | $dm^3/min$ | 54,3 |

The course of the laboratory test was based on the operational measurement of the hydraulic system of the agricultural wheeled tractor ZETOR FORTERA 114 41. We determine the procedure for evaluating the high-performance product line of the QHD 17 gear pumps with the Vickers test method. Test conditions defined by the Vickers test method are stated in Table 3. Tested work fluid is the ecological transmission-hydraulic oil MOL Farm NH Ultra which properties are from the manufacturer and are stated in Table 1.

**Table 3.** Test conditions according to the test method.

| Parameter | Unit | Value |
|---|---|---|
| Rotations | 1/min | 2350–2400 |
| Loading pressure | MPa | 20,5–21 |
| Inlet pressure | MPa | max. 0,15 |
| Outlet temperature | °C | 90–96 |

**Table 4.** Basic properties of the tested hydraulic fluid Mol Farm NH Ultra.

| Characteristic | Unit | Value |
|---|---|---|
| Density at 15 °C | kg/m | 875 |
| Kinematic viscosity at 40 °C | $mm^2/s$ | 64,2 |
| Kinematic viscosity at 100 °C | $mm^2/s$ | 10,9 |
| Viscosity index | - | 162 |
| Pour point | °C | -36 |
| Flash point in open container | °C | 210 |

The Vickers test is determined to be 50 hours without interruption, while the standard determines the different heating phases and the flow rate measurements are taken in 5-hour intervals. In our laboratory conditions, we have adjusted this test to 100 hours by simulating a pressure load that will be intermittent after 10 hours and the flow characteristics of the pump will be measured and recorded in 25-hour intervals. The change in fluid flow for gear pumps is determined by the Vickers test method only by a maximum of 3,75%. Methodological procedure for sampling of hydraulic fluid tested is governed by the standard STN 65 6207 (Hydraulic oils and liquids. Sampling for determination of mechanical impurities content.). In order to assess the suitability of the hydraulic fluid tested in the hydraulic circuits of agricultural, forestry and handling equipment, it is necessary to know the input parameters of the system in which the hydraulic fluid will be tested, i.e. the technical parameters of the pump (Table 2).

# 3 Results

## 3.1 Flow characteristics of the pump overtime

In Fig. 1 is a graphical comparison of the flow rate variation of the hydraulic pump after 500 hours of operating load simulation under laboratory conditions where the flow rate variation is depicted depending on the speed range from 500 rpm to 2750 rpm.

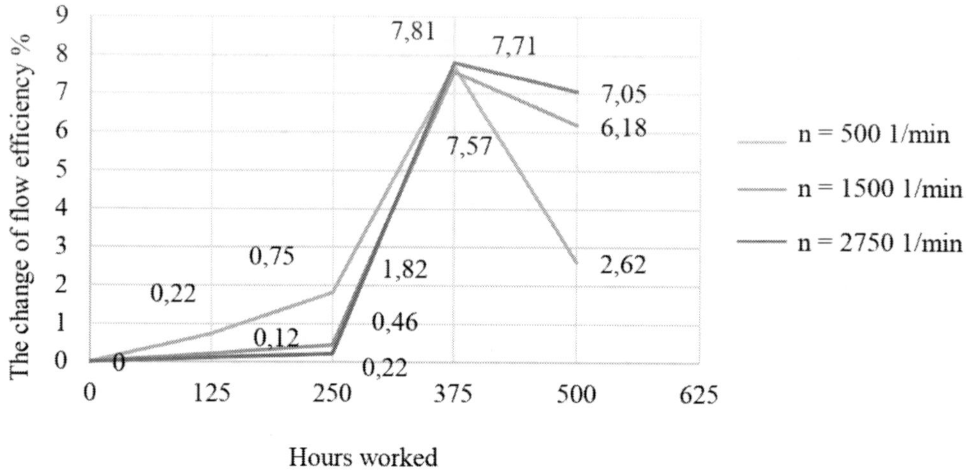


**Fig. 1.** Comparison of the change in flow efficiency from based on hours worked.

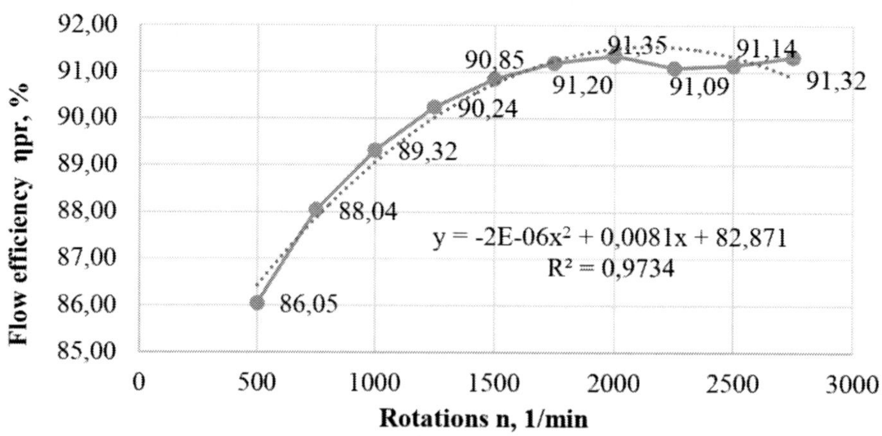


**Fig. 2.** Flow efficiency of the hydraulic pump at 0 hours – Vickers test method.

Fig. 2 shows the flow efficiency of the pump after 500 hours worked, depending on the minimum speed, from 500 rpm to 2750 rpm. Flow efficiency at 500 rpm reached 86,05 % and at 2750 rpm reached 91,32%. The greatest increase in flow efficiency as a function of speed is between 500 rpm and 1500 rpm (4,8%), after exceeding the rated speed of 1500 rpm to the value of 2750 rpm the flow efficiency increased (0,47%).

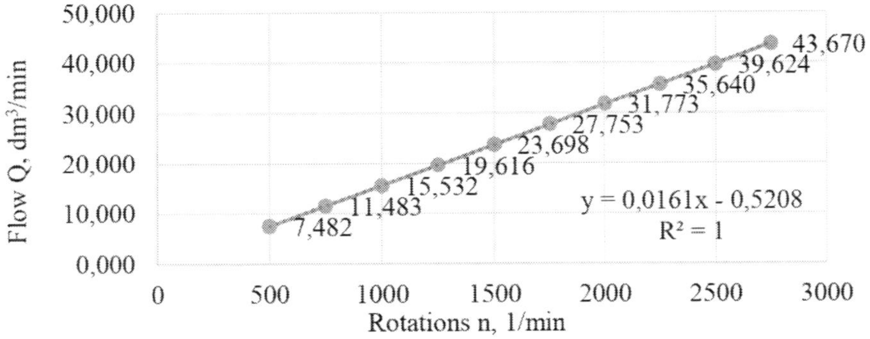


**Fig. 3.** Flow characteristic of the hydraulic pump at the start of the test Vickers.

Fig. 3 shows the flow characteristic of the pump after 500 hours worked, where we started the measurement with the Vickers test method depending on the minimum speed from 500 rpm to 2750 rpm. The flow rate at 500 rpm reached 7,48 $dm^3$/min and at 2750 rpm it reached 43,67 $dm^3$/min. The increase in the flow characteristic has a linear course as a function of speed, with a value of 36,19 dm3/min between 500 rpm and 2750 rpm. Table 5 shows the flow characteristics of the pump after 25 hours worked in accordance with the Vickers test method in dependence on a minimum speed of 500 rpm to 2750 rpm.

**Table 5.** Basic statistical evaluation of the flow characteristics of the hydraulic pump after 25 hours – test Vickers.

| Flow Q at rotations n 1/min | Average | Median | Modus | Min. value | Max. value | Spread | Stand. deviation | Var. coef. | Flow efficiency |
|---|---|---|---|---|---|---|---|---|---|
| | $\bar{x}$ | Me | Mo | $x_{min}$ | $x_{max}$ | $\sigma^2$ | $\sigma$ | $V_k$ | % |
| 500 | 8,030 | 8,040 | 8,020 | 7,930 | 8,160 | 0,001 | 0,038 | 0,471 | 92,35 |
| 750 | 12,488 | 12,490 | 12,530 | 12,270 | 12,670 | 0,006 | 0,076 | 0,609 | 95,75 |
| 1000 | 16,883 | 16,880 | 16,880 | 16,750 | 17,010 | 0,003 | 0,056 | 0,331 | 97,08 |
| 1250 | 21,316 | 21,310 | 21,290 | 21,170 | 21,450 | 0,003 | 0,055 | 0,257 | 98,06 |
| 1500 | 25,776 | 25,770 | 25,770 | 25,630 | 25,940 | 0,004 | 0,064 | 0,248 | 98,82 |
| 1750 | 30,197 | 30,180 | 30,140 | 30,060 | 30,370 | 0,005 | 0,073 | 0,242 | 99,22 |
| 2000 | 34,566 | 34,570 | 34,650 | 34,390 | 34,720 | 0,006 | 0,078 | 0,226 | 99,38 |
| 2250 | 38,844 | 38,850 | 38,850 | 38,470 | 39,110 | 0,021 | 0,143 | 0,369 | 99,28 |
| 2500 | 43,095 | 43,110 | 42,920 | 42,800 | 43,430 | 0,026 | 0,161 | 0,373 | 99,12 |
| 2750 | 47,366 | 47,360 | 47,310 | 47,050 | 47,670 | 0,020 | 0,143 | 0,302 | 99,05 |

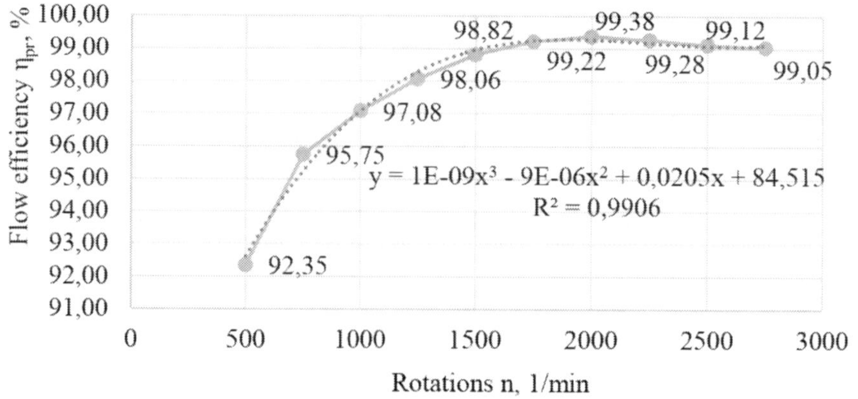


**Fig. 4.** Flow efficiency of the hydraulic pump after 25 hours – test Vickers.

Table 4 shows the flow efficiency of hydraulic pump after 25 hours, between 500 rpm to 2750 rpm. Flow efficiency at 500 rpm was 92,35% and at 2750 rpm the flow efficiency was 99,05%. The biggest increase in flow efficiency based on rated speed was between 500 rpm and 1500 rpm, at 6,47%, after exceeding the rated speed of 1500 rpm to the value of speed 2750 rpm, the flow efficiency increased (0,23%).

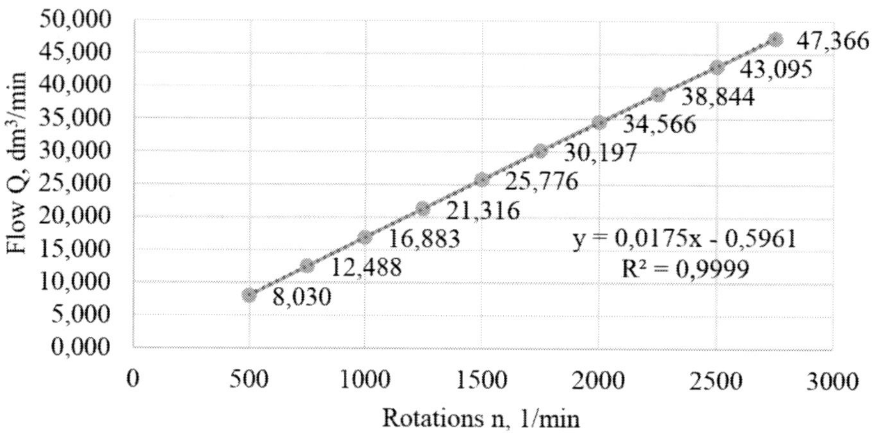


**Fig. 5.** Flow efficiency of the hydraulic pump after 25 hours – test Vickers.

Table 5 shows the flow rate characteristic of the pump after 25 hours depending on the minimum speed of 500 rpm to 2750 rpm. The flow rate at 500 rpm reached 8,03 $dm^3$/min and at 2750 rpm it reached 47,37 $dm^3$/min. The increase in the flow characteristic has a linear course as a function of speed, with a value of 39,34 $dm^3$/min between 500 rpm and 2750 rpm. Comparing the flow efficiency of the pump after working for 25 hours, according to the Vickers test method, the flow rate shows an increase in flow efficiency of 8,77% at a rated speed of 1500 rpm. The increase in flow efficiency is attributed to changes in the physico-chemical properties of the liquid and changes in laboratory test conditions. The working temperature of the monitored liquid changes from 60 °C during a 500-hour test to 90–96 °C, the operating load changes from the maximum pressure of 14,11 MPa during a 500-hour test to a value of 21 MPa during a 100-hour test. The Vickers test has changed the kinematic viscosity, which is attributed to an increase in flow efficiency.

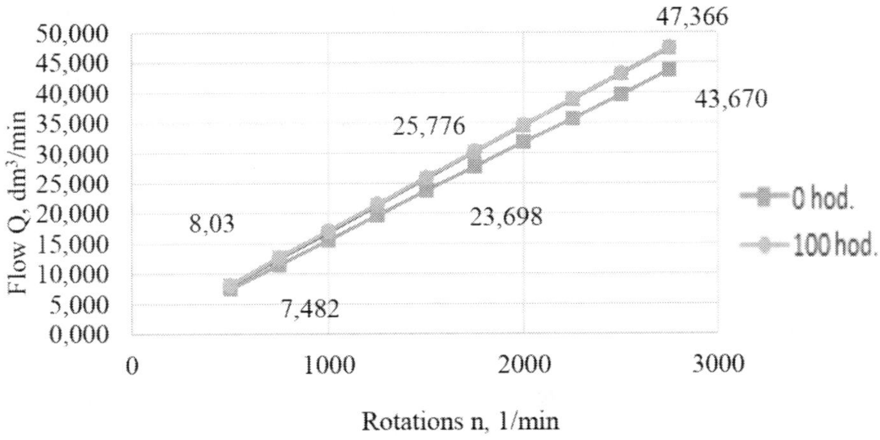


**Fig. 6.** Comparing the flow characteristics of pump – test Vickers.

Fig. 6 shows a graphical comparison of the flow characteristics of the hydraulic pump after each hour of working in accordance with the Vickers test conditions versus a minimum speed of 500 rpm to 2750 rpm. When comparing the flow rate of the pump after working for 25 hours, according to the Vickers test method, the flow chart shows an increase in flow rate of 8,77% at a rated speed of 1500 rpm. Given the established methodology and the Vickers test method for gear pumps, a flow rate variation of 3,75% is permissible, suggesting that the biodegradable universal fluid does not comply with the Vickers test, after the first 25 hours of laboratory testing under standardized conditions Vickers test.

## Conclusion

According to [15] and [16], the hydrostatic pressure transfer system and the use of the hydrodynamic principle requires a perfect medium, therefore the hydraulic fluid must meet all the conditions encountered in the operation of hydraulic systems. Determination of the state and transfer of thermal and oxidative degradation of the oil is the most important criterion for oil life [5]. While using the biodegradable universal fluid, the flow rate of the hydrogenator after 500 hours worked reached 7,48 $dm^3$/min at 500 rpm and at 2750 rpm it reached 43,67 $dm^3$/min, an increase of 36,19 $dm^3$/min. After a subsequent 100-hour Vickers test, the hydraulic pump flow rate at 500 rpm reached 8,14 $dm^3$/min and at 2750 rpm it reached 47,40 $dm^3$/min, an increase of 39,25 $dm^3$/min. The flow characteristics were almost linear over the entire duration of the laboratory test. The stated flow characteristics of the hydraulic pump correspond with the measured result according to [17], during the verifying measurement of a proposed laboratory test facility for testing hydraulic pumps and transmission-hydraulic fluids. When comparing the flow characteristic of the pump after working for 0 to 25 hours, according to the Vickers test method, an increase in flow characteristic of 8,77% at a rated speed of 1500 rpm was observed. In view of the established methodology and the evaluation of the Vickers test method for gear pumps, it is permissible to change the flow rate by 3,75%, indicating that the biodegradable universal liquid does not comply with the Vickers test after the first 25 hours of laboratory testing conditions according to the Vickers test.

This work was supposed by project VEGA 1/0155/18 „Applied research of the use ecological energy carriers in agriculture, forestry and transport technology."

This work was supposed by project KEGA 028SPU–4/2019 „Practical utilization of design and testing knowledge of transmission systems of hydraulic mechanisms of mobile agriculture and forestry machinery."
This work was supposed by project APVV SK-PL-18-0041 „The development of scientific cooperation in the study of the effect of biofuels in road transport, including environmental impact."

## References

1. J. Kosiba, Z. Tkáč, Ľ. Hujo, B. Stančík, I. Štulajter, RAE, **59**, 27 (2013)
2. Z. Tkáč, Š. Čorňák, V. Cviklovič, J. Kosiba, J. Glos, J. Jablonický, R. Bernát, ATA, **20**, 42 (2017)
3. Ľ. Hujo, Š. Čorňák, Z. Tkáč, M. Jánošová, TAE, 183 (2019)
4. I. Kuric, I. Zajačko, M. Cisár, Adv. Sci. Technol. Res. J., **10**, 59 (2016)
5. M. Puškár, A. Jahnátek, I. Kuric, J. Kádárová, M. Kopas, M. Šoltésová, Air Qual. Atmos. Health , **12**, 855 (2019)
6. J. Jablonický, Z. Tkáč, R. Majdan, D. Uhrinová, Ľ. Hujo, V. Vozárová, *Hodnotenie vlastnostní biopalív a biomazív* (2012)
7. J. Tulík, Ľ. Hujo, J. Kosiba, J. Jablonický, M. Jánošová, RAE, **63**, 1 (2017)
8. M. Kučera, Z. Aleš, Z. Ivandić, Ľ. Hujo, JCEA, **14**, 1592 (2013)
9. M. Košinár, I. Kuric, Adv. Sci. Technol., 6, 115 (2011)
10. M. Zastempowski, J. Agr. Eng. Res., **58**, 188 (2013)
11. J. Kosiba, Š. Čorňák, J. Glos, J. Jablonický, V. Vozárová, A. Petrovič, J. Csillag, JAR, **14**, 1626 (2016)
12. M. Puškár, T. Brestovič, N. Jasminská, Int. J. Veh. Des., **67**, 63 (2015)
13. B. Kopiláková, J. Turza, Ľ. Hujo, J. Kosiba, Tribology in Industry, **39**, 129 (2017)
14. Š. Čorňák, ERDev, 2048 (2018)
15. Ľ. Hujo, P. Kangalov, J. Kosiba, *Laboratory test devices for evaluating the lifetime of tractor hydraulic components* (2015)
16. M. Simikić, N. Dedović, L. Savin, M. Tomić, O. Ponjičan, Soil Till. Res., **141**, 32 (2014)
17. Ľ. Hujo, Z. Tkáč, J. Tulík, J. Kosiba, D. Uhrinová, M. Jánošová, RAE, **62**, 24 (2016)

# Impact of the content of alcohol in petroleum on the level of an unsupercharged engine's noise

*Jerzy* Kaszkowiak[1,*], *Lubomir* Hujo[2], and *Juraj* Jablonicky[2]

[1]UTP University of Science and Technology, 85-796 Bydgoszcz, ul. Kaliskiego 7 Poland
[2]Slovak University of Agriculture 949 76 Nitra Tr. A. Hlinku 2 Slovakia

**Abstract.** The results of the studies conducted on the chassis dynamometer for a vehicle powered with a spark-ignition engine are presented in the study. The tested vehicle was equipped with an unsupercharged engine. There was tested the noise level of the engine powered with ethyl alcohol at variable settings of the fuel injection. In the course of the tests, the dose of fuel was increased within the range from the manufacturer's settings respectively by: 5%, 10%, 20% 30% and 50% of the nominal dose, and the ignition advance angle with reference to manufacturer's settings was increased respectively by $0^0$ and $3^0$ . It was found, that the noise level for the engine powered with ethanol decreases together with the increase of the fuel's dose, (up to about 110% of the nominal dose), and then increases. At the same time, the increase of the noise level for the increased value of the ignition advance angle was found.

## 1 Introduction

Because of the environment protection necessity, looking for alternative energy sources for machines and vehicles, constitutes the subject matter of interest of numerous research centres, as well as companies dealing with production of engines. One of the manners to reduce arduousness for the environment, are electric drives or use of alternative fuels comprising components originating from renewable sources. These fuels should allow for operation of engines without the negative impact on their durability and ensure comparable performance just as in case of powering with traditional fuels [1]. In the conditions of the Central and Western Europe, popular is production of ethyl alcohol from organic biomass. These may be wastes from forest or agricultural production, as well as municipal wastes [2,3], or plants cultivated specially as a raw material for biofuels' production. Important are also possibilities and costs of ethanol obtaining. In the costs of production, a considerable share, apart from the prices of raw material, there have energy outlays and raw-material's processing [4].

For engines with self-ignition, bioesters are such an additive [5]. In case of diesel engines with spark ignition, the alternative for fuels of connate origin, are among the other alcohols. They may be combusted independently or – what happens most often – as an additive to petroleum. For that purpose, both the ethyl alcohol, as well as other alcohols are used, in particular the methyl and butyl alcohol [6].

Alcohols, as fuels, are most often characterised by a lower calorific value, but also a higher-octane number. Moreover, for the course of the combustion process, they require lower volume of oxygen than the fuels of connate origin.

Properties of the selected alcohols are presented in the table 1. For most of the currently used engines, use of slight alcohol additives, does not cause any negative results. At the territory of the European Union, in most of the countries the additive of ethyl alcohol in the volume of 5% to petroleum is used. Such an additive, apart from decreasing of the connate

* Corresponding author: kaszk@utp.edu.pl

fuels resources' consumption, in most cases influences neither the composition of the combustion gases' composition nor the power and the engine's turning moment. According to different studies, only the additive of alcohol in the volume above 30%, results in the decrease of the content of harmful compounds in exhaust gases. [7, 8,9].

**Table 1.** Selected properties of petroleum, ethanol and methanol.

| Name/parameter | Density in 15°C [kg·m-3] | Research octane number | Motor octane number | Boiling point [°C] | Calorific value [MJ·kg-1] | Heat of evaporation [kJ·kg-1] |
|---|---|---|---|---|---|---|
| Methanol | 795,6 | 107 | 91 | 64,7 | 19,9 | 1100 |
| Ethanol | 793,2 | 108 | 92 | 78,3 | 26,8 | 910 |
| Petroleum 95 | 720-775 | 95 | min. 85 | 215 | 41,0 | 380 |

Limitation of connate fuels' resources and improvement of the exhaust gases' composition occurring as a result of ethyl alcohol combustion, results also in changes in the noise level emitted by engines. It is very often connected with the lower noise level connected with the power-supply system (reduced flow of the air), as well as smaller volume and speed of the combustion gases outflow.

## 2 Materials and methods

In the tests, there was used the vehicle Fiat Panda, powered with 8-valves engine with spark ignition and multiport fuel injection. The engine at manufacturer's settings and powering with petroleum E 5, reached the power 61,8 KM (47,2 kW), and the turning moment 100,5 Nm. The course of power and turning moment at manufacturer's settings, is presented in the figure 1.

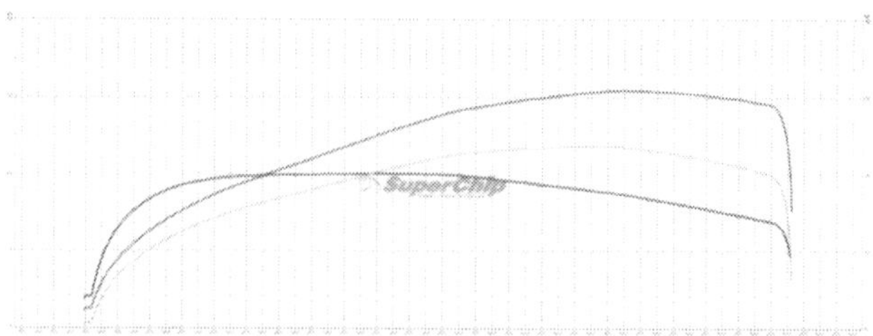

**Fig. 1.** Course of the power and turning moment in the tested engine at manufacturer's settings and powering with fuel E5.

Prior to measurements, in the engine the engine oil and oil filter, air and fuel filters were replaced. The vehicle was immobilized on rollers. Powering with fuel was connected passing over the original fuel tank. A special tank of the capacity of 5 dm$^3$ was used for that purposes. The engine was heated up to the moment of reaching by the cooling liquid the temperature 75°C. In the room of the test house, at the time of the tests conducting, there prevailed the temperature of 15°C and the pressure was 100,4 kPa. The view of the vehicle at the testing stand with the connected external tank, is presented in figure 2.

**Fig. 2**. The view of the vehicle at the testing stand with the connected external tank.

In the vehicle there was replaced the factory's control computer into an universal one, with the option of free settings' programming. The manufacturer's software was installed, and the course of the power and moment was checked again. The programming computer was connected. View of the vehicle and the programming computer at the test stand is presented in figure 3.

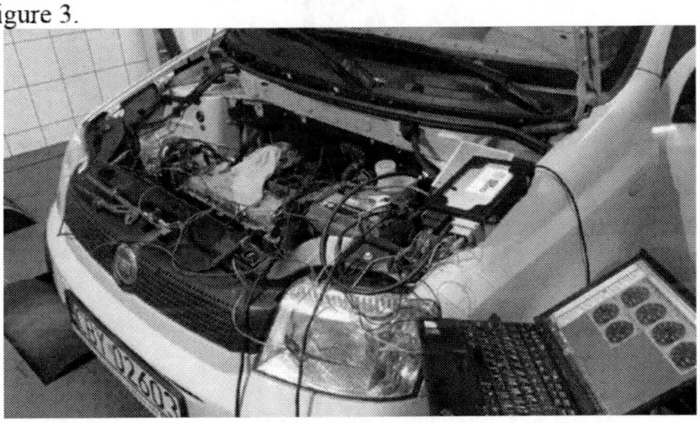

**Fig. 3**. View of the vehicle at the test stand with connected modifying computer.

The noise level was measured with the use of the integral noise level measuring device HD 2010UC. The measuring device has been assembled at the distance of 1 m from the engine, at the front of the vehicle. The measurements were conducted for alternately changed engine's settings. The view of the measuring device is presented in figure 4. In the control computer's software, the reaction to readings of the knock sensor has been turned off. The diagram of the engine's settings is presented in the table 2.

**Table 2.** Diagram of the engine's settings.

| Fuel dose | Ignition lead | |
|---|---|---|
| Manufacturer's setting | manufacturer's | - |

| | | |
|---|---|---|
| Fuel dose 105% | manufacturer's | $+3^0$ |
| Fuel dose 110% | manufacturer's | $+3^0$ |
| Fuel dose 120% | manufacturer's | $+3^0$ |
| Fuel dose 130% | manufacturer's | $+3^0$ |
| Fuel dose 150% | manufacturer's | $+3^0$ |

Each measurement was repeated three times. At the time of the measurements' conducting, the ambient temperature and atmospheric pressure was controlled.

**Fig. 4.** View of the noise level measuring device HD 2010UC.

The obtained results were subject to statistical analysis. Mean values are presented at the diagrams 5, 6 and 7.

## 4 Analysis of the results and conclusions

The results obtained as the effect of the conducted researches, were subject to statistical analysis with the use of the STATISTICA programme. Occurrence of statistically significant differences were disclosed both depending on the fuel dose, as well as the ignition advance angle. The mean values of the noise level for the manufacturer's setting of the ignition advance angle is presented in diagram 5.

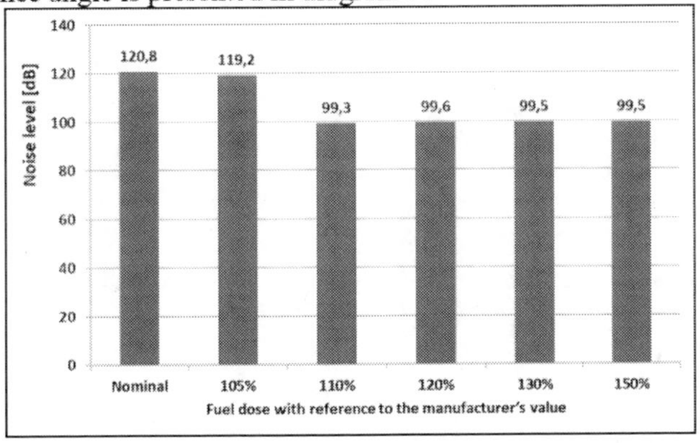


**Fig. 5.** Noise level for the manufacturer's value of the ignition advance at the increasing ethanol dose.

For the manufacturer's and increased by 5% doses, there has been found the statistically essential noise level. At the dose's increase by more than 10%, the noise level lowered almost by 20 dB, not changing significantly at the further dose's increase. Based on the observations conducted during the researches it should be assumed, that lower ethanol doses were combusted in a detonation manner. It resulted from the stoichiometric air excess at the time of the ethyl alcohol combustion, having a smaller demand for air for combustion.

For combustion of fuel at the increased by $3^0$ ignition advance angle, the increase of the noise level was found. At the nominal dose, increased by 0 5% and 10%, the noise level did not differ between them significantly statistical. At the fuel doses increased by 20, 30 and 50%, there was found statistically significant increase of the noise level up to the value comparable with the measurements for the nominal setting of the ignition advance angle and lower fuel doses.

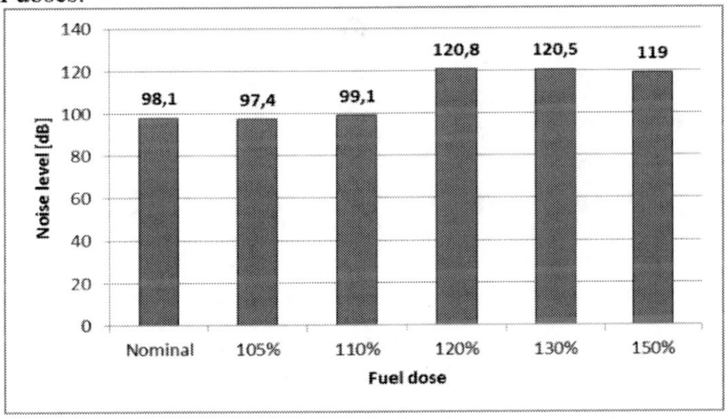


**Fig. 6.** Noise level for the manufacturer's value of the ignition advance at the increasing ethanol dose.

The mean noise level emitted during the engine's operation for the nominal ignition advance angle, had the value by 3 dB lower, than for the angle increased by $3^0$. The dependence is presented in diagram 7.

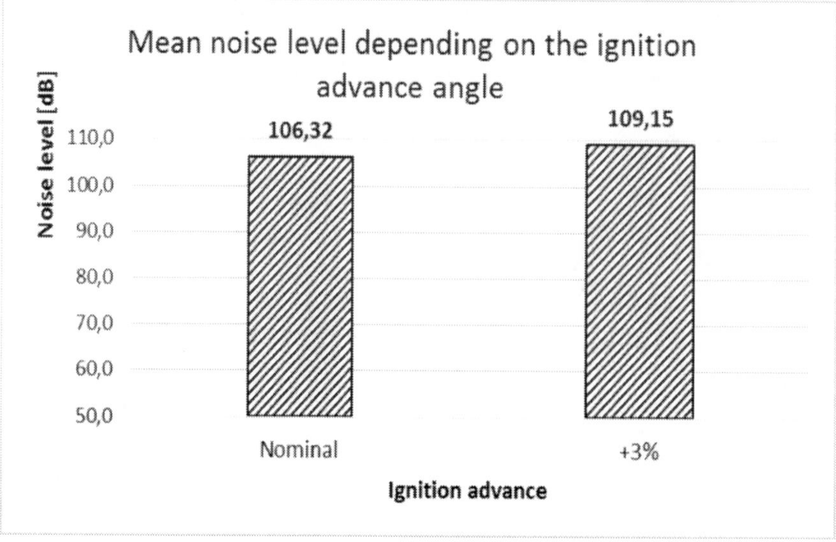


**Fig. 7.** Mean noise level for the nominal and increased value of ignition advance.

At the increasing dose of fuel, the mean noise level reached the lowest value at the fuel dose increased by 10% as compared to the nominal value. Both for the lower and higher volume of fuel, the noise level has essentially statistical higher value.

Course of the values of the noise level is presented in figure 8.

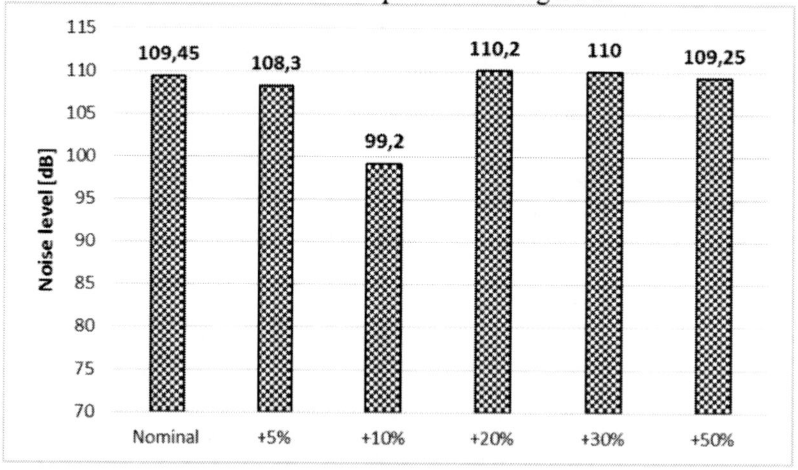


**Fig. 8.** Mean noise level depending on the volume of the fuel dose

Value of the noise level for the fuel doses for 20, 30 and 50% higher than the nominal ones, did not differ essentially statistical between themselves.

The value of noise at the increasing fuel dose, both at the nominal ignition advance angle and at the increased one was compared. It was found, that the lowest value the noise level was reached at the dose increased by at least 10%. It did not significantly statistically increase at the doses higher by 20, 30 and 50% than the nominal one. However, for the ignition advance angle increased by $3^0$ with reference to the nominal one, the doses higher by 20, 30 and 50% resulted in the noise level's increase. The measured values for 20, 30 and 50%, did not essentially statistically differ between themselves.

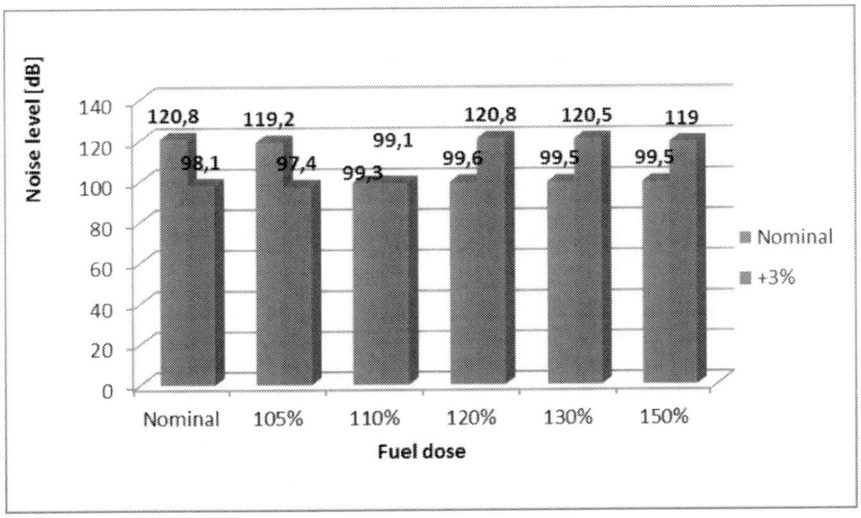


**Fig. 9.** Mean level of noise for the nominal and increased value of ignition advance

As mentioned already, in the methodology of the studies, in the vehicle the control of the knock sensor's control was switched off in the engine. The increase of the noise level at low fuel doses, was caused by detonation fuel combustion in the excessive air volume. In case of the fuel dose's increase, the noise level originally dropped and then raised, what could have been caused by the increased volume of combustion gases and their increased outflow speed.

## References

1. K. Baczewski, T. Kałdoński *Paliwa do silników o zapłonie samoczynnym* (2008)
2. I. Niedziółka, M. Szymanek, A. Zuchniarz Inż. Roln. **95** (2007)
3. M. Zastempowski, S. Borowski, J. Kaszkowiak Conference Proceeding - 5th International Conference, TAE: pp. 673-676 (2013)
4. S. Borowski, L.A. Knopik, M.M. Markiewicz, A. Brzostek, Proceeding of 6th International Conference on Trends in Agricultural Engineering pp 845-851, (2016)
5. J. Jablonicky, Z. Tkac, at al. *Hodnotenie vlastnosti biopaliv a biomaziv* (Nitra 2012)
6. A.Yasar, Metalurgija **49** pp 335-338 (2010)
7. G. Evangelos, N. Giakoumis, C. Dimitrios, S. Rakopoulos, D. A rev. Renew. and Sustain. En. Rev. **54** pp1099–1113, (2016)
8. Z. Zbytek, Z. Dach at.al. *International Conference on Chemical and Biological Sciences* Amsterdam pp 644-648 (2016)
9. J. Wasilewski, P. Krzaczek Przem. Chem. **93** pp 343-346, (2014)

# Methods for Vibration Reduction in Enclosed Electronic Packages

*Igor* Kovtun[1,*], *Juliy* Boiko[2], *Svitlana* Petrashchuk [1], *Tomasz* Kałaczyński [3]

[1]Technologies and Design Faculty in Khmelnytsky National University, Department of Art and Technical Graphics, 29016, 11, Instytuts'ka str., Khmelnitsky, Ukraine
[2]Programming and Computer, Telecommunication Systems Faculty in Khmelnytsky National University, Department of Telecommunication and Radio Engineering, 29016, 11, Instytuts'ka str., Khmelnitsky, Ukraine
[3]Faculty of Mechanical Engineering UTP in Bydgoszcz, Department of Vehicle Engineering, Al. prof. S.Kaliskiego 7, 85-796 Bydgoszcz, Poland

**Abstract.** Excitation and transmission of vibration inside electronic packages are strongly connected with mechanical characteristics of their enclosure, which are in the objective for the current research aimed at developing methods and means to reduce vibration inside the enclosure case. The main idea of the introduced methods is in creating elastic and dissipative mechanical joints between structural elements, which appear to be the links for vibration transmission, inside and outside of the electronic package enclosure. The effectiveness of developed methods is proved by experimental verification and theoretical foundation.

## 1 Introduction

Vibration analysis of enclosed electronic packages represented in [1-3] disclosed that printed circuit boards (PCB) installed inside electronic packages may receive resonant oscillations, which amplitudes increase manifold and cause breakages of parts and components or deviations of electronic parameters. Excitation and transmission of vibration inside electronic packages are strongly connected with mechanical characteristics of their enclosure, which are in the objective for the current research aimed at developing methods and means to reduce vibration inside the enclosure case.

Despite variety of vibration isolators [4-7] the reliable and simple wideband isolator applicable to electronic packages has yet to be created. Now the close attention is paid to damping methods having inherent drawbacks though. For an instance, the widely used damping by encapsulation of electronic packages with vibration absorbing materials makes the package not maintainable and causes internal stress and hence damages to sealed components and sealant when subjected to thermal impacts as it was demonstrated in [8,9].

The previous research [1-3] verified that developing methods for protection against vibration and avoiding resonances is achievable by complex using of several methods unlike improving the single one. However the leading role in the task belongs to damping.

## 2 Mechanical characteristics of enclosure case

Enclosure case of electronic package represents one of the links in vibration transmission to PCBs installed inside. Hence varying its mechanical characteristics, which are elastic and dissipative parameters, such as stiffness, are expected to effect vibration transmission in purpose to reduce it. In this purpose the enclosure case was varied by changing stiffness of the bearing walls and introducing three modifications: 1) enclosure case with walls of 1 mm

* Corresponding author: dr.igorkovtun@gmail.com

thick aluminum; 2) enclosure case with walls from damping material (1 mm thick fiberglass); 3) enclosure case with walls removed.

The tested objects were represented by two types of printed circuit boards $PCB_1$ and $PCB_2$, which differed in elastic and inertial characteristics: stiffness $C_1 < C_2$ and mass $m_1 < m_2$. Vibration analysis for PCBs was conducted by two parameters measured on their surface: vibration amplitude and strain. Vibration sensor (V) and strain gauges (S) were attached to PCBs as shown in fig. 1.

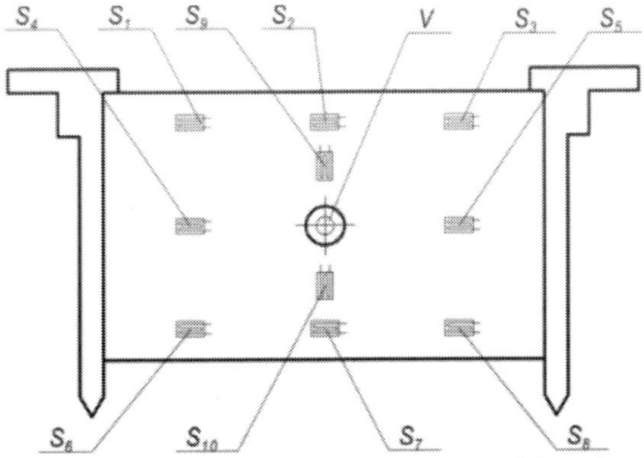


**Fig. 1.** Layout of vibration sensor and strain gauges on printed circuit boards.

The generated frequency and amplitude of testing vibration was controlled by the shaker VEDS-200A. In the experimental setup the tested electronic package was installed on the shaker so that PCBs inside were in horizontal position and direction of generated vibration coincided with their minimal stiffness; in such position oscillation amplitude of PCBs is maximal. Vibration tests were performed by method of fixed resonance frequencies, which had been identified by trial tests conducted before by using varying frequency method in diapason from 0 to 500 Hz; important that generated vibration acceleration was maintained stable. Noteworthy is that tests were conducted in frequencies under 100 Hz, since the higher diapason (100-5000 Hz) had no excitation of oscillations in the tested objects observed. Tests were performed for all modifications of cases. Tightening torque of bolts mounting tested case onto shaker table was maintained equal to 2.8 N·m in order to eliminate its effect to vibration transmission.

Results of experiments were represented by amplitude frequency responses obtained for two types of PCBs in three modifications of the cases (fig. 2). To unify measured values they were converted into relative values introduced as vibration transmission:

$$w = a_b / a_0 \tag{1}$$

where: $a_b$ – vibration acceleration measured on printed circuit boards, m/s$^2$; $a_0$ – vibration acceleration generated by the shaker, m/s$^2$.

Obtained amplitude frequency responses of vibration transmission revealed that different enclosure case modifications demonstrate different vibration transmission to PCBs under the same level of applied dynamic force. Such difference is explained by their different damping level and confirms the statement that structural damping surpasses the damping produced by internal friction in material of structure parts.

The issue of the research was that decreasing case stiffness, in particular bearing walls, increases vibration transmission from the case supports to printed circuit boards. $w$ in case with walls removed is 2-4 times higher, and in case with walls from fiberglass – 1.5-2 times higher in comparison with regular case with aluminum walls.

At the same time fig. 2 demonstrated negligible changes in resonant frequencies for both types of PCBs in different modification of enclosure case: type 2 increased resonance frequency by 1-1.5 Hz and type 3 – by 1.5-2.5 with respect to type 1. Thus, offered changes in design of enclosure case were not effective for detuning from resonance.

Analysis of vibration acceleration measured for two types of PCBs revealed that decreasing elastic and inertial characteristics, such as mass and stiffness, increases amplitude response at resonance frequencies. This fact attracts closer attention to $PCB_1$.

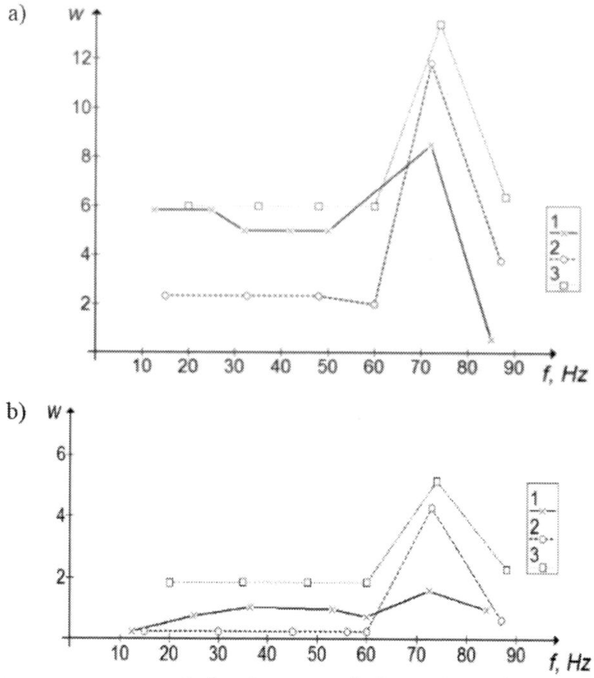


**Fig. 2.** Amplitude frequency response of vibration transmission to $PCB_1$ (a) and $PCB_2$ (b) in types of cases: 1, 2, 3.

In order to estimate level of stress caused by vibration tests in $PCB_1$ strain values measured at resonant frequencies were recalculated [10,11] to stress values by Hooke's law for linear deformation. Mean values of stress were calculated for PCB substrate material (fiberglass) whose elasticity module is $E = 0.2 \cdot 10^5$ MPa.

The maximal vibration acceleration and stress were observed at frequency 72 Hz, which is the natural frequency for the researched PCB, whilst at other resonances, which were produced by other parts of enclosure case, these values are much lower. Those additional resonances depended on the stiffness of the case walls, which also effects to vibration transmission from the case to PCBs. Vibration acceleration and stress observed at additional resonances were negligibly low. The highest stress $\sigma = 30$ MPa was measured in PCBs installed in enclosure case without side walls. This stress was 3 times higher than that of the case with aluminum walls. Vibration transmission reached almost 9 times inside the case with aluminum walls and almost 13 times for the case without side walls.

Another experiment revealed influence of tightening torque of bolts mounting case to the shaker's platform on vibration transmission from the shaker to PCBs – growing tightening torque reduces vibration transmission to PCBs. This issue confirms the statement made above, that reduction of case stiffness increases vibration transmission to PCBs. In the test the case of type 1 with $PCB_1$ installed inside was subjected to vibration at the resonant frequency 72 Hz and vibration acceleration 50 m/s$^2$ along with gradual changing tightening torque of bolts.

Thus, high stiffness is required for building enclosure case even from damping materials in order to avoid additional resonances capable to increase vibration transmission inside the enclosure. Among the cases whose design differed in stiffness and other elastic, inertial and dissipation parameters the case with aluminum walls (type 1) was considered as the most effective for resonant oscillation reduction. Nevertheless, vibration transmission to PCBs in this case remains yet unacceptably high. Therefore searching effective methods reducing vibration transmission and excitation was considered actual for the next research.

## 3 Elastic and dissipative joints of enclosure case

$PCB_1$ installed in case of type 3 was in objective in the next research. Case without side walls was selected in purpose to eliminate effect that walls produce to amplitude frequency response. The main goal of the research was developing methods for reduction of vibration transmission and excitation in PCBs, and the main idea consisted in building and studying elastic and dissipative joints in two options: 1) outside of enclosure case – between case and source of vibration; 2) inside of enclosure case – between its bearing parts.

The first option was introduced by two variants for installation of enclosure case on the shaker: 1) rigid installation, when enclosure case is firmly fixed on the shaker; 2) flexible installation, when enclosure case is set on the shaker through rubber spacers.

The second option was introduced by building three variants of mechanical joints into PCB installation: 1) banding; 2) inserting bounding rail; 3) inserting bounding rail through rubber spacers. All variants were tested in flexible installation of the package on the shaker.

Banding of PCBs was executed by using 4 mm thick cable passed through technological holes in all PCBs inside the case (fig. 3). Banding was applied with reasonable tension to provide necessary stiffness of mechanical joint.

Rails made from composite material were profiled with recesses cut at one side with spaces equal to installation steps for PCBs in the case (fig. 3). Another variant of using rails was inserting them through the rubber spacers.

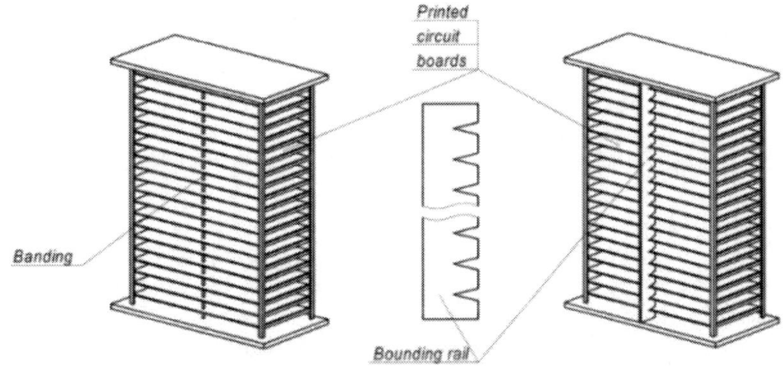


**Fig. 3.** Parts of elastic and dissipation joints inside enclosure case.

Reading vibration was provided by vibration sensor IS567A-III attached to PCBs in their center. The second vibration sensor was attached to the shaker's platform. Procedure of vibration tests was identical to the previous chapter. Electronic package was subjected to vibrations at resonance frequencies, which had been detected before. The shaker generated vibration acceleration of 10 m/s$^2$ in frequency diapason 20-40 Hz and 50 m/s$^2$ in frequency diapason 50-2000 Hz. Every test was repeated three times to reduce measuring errors. Values of vibration transmission measured on PCBs in all variants of elastic and dissipative joints are represented in fig. 4. Tightening torque of fixing bolts was maintained equal to 1.4 N·m.

Figure 4 indicates natural frequency of $PCB_1$ around 60 Hz, what was then verified by striking method. All variants tested in the research demonstrated their effectiveness to reduce vibration transmission to $PCB_1$ although its natural frequency was not noticeably changed.

The highest effectiveness was achieved by inserting bounding rail, which more than 5 times reduced vibration transmission. Application of bounding rail has its obvious advantage as its usage is complimentary and does not change production technology. Close result was reached by inserting bounding rail with rubber spacers.

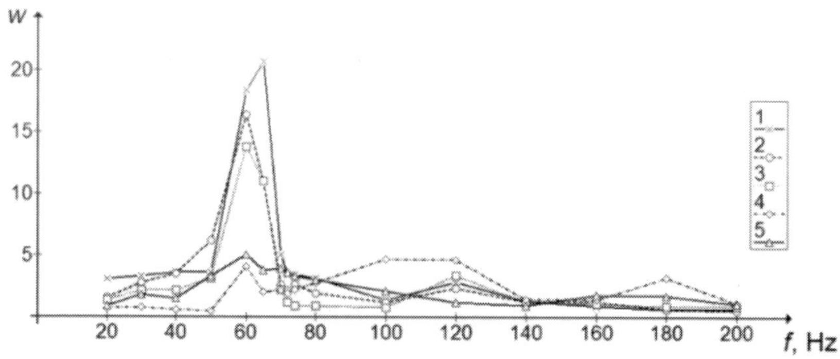


**Fig. 4.** Amplitude and frequency response of vibration transmission for enclosure case installations: 1 – rigid; 2 – flexible; and PCB installations: 3 – banding; 4 – inserting bounding rail; 5 – inserting bounding rail through rubber spacers.

# 4 Installation of printed circuit boards

Since vibration of PCB in resonance exceeds manifold vibration of its enclosure the next research was aimed at creation of elastic and dissipative joints in fixtures of PCBs inside the enclosure case. To perform variety of the fixtures in tests the vise appliance was built. PCBs fixed in vice were set on the shaker.

The following modifications represented PCBs fixtures as:

1. regular plastic fixture (fig. 5, a);
2. vibration isolating fabric tape (fig. 5, b);
3. vibration isolating rubber tape (fig. 5, b);
4. regular plastic fixture with foam rubber inserts;
5. special fixture pasted over by rubber tape;
6. vibration isolating fabric tape with dry friction damper (fig. 5, c).

The test procedure was identical to the previous research. Vibration accelerations were read from two piezoelectric sensors: one attached to the center of PCB and another – to the shaker's platform. Amplitude and frequency response of vibration transmission shown in fig. 5, d indicated the fixture of PCB on the vibration isolating fabric tape with dry friction damper as the most effective protection method against vibration.

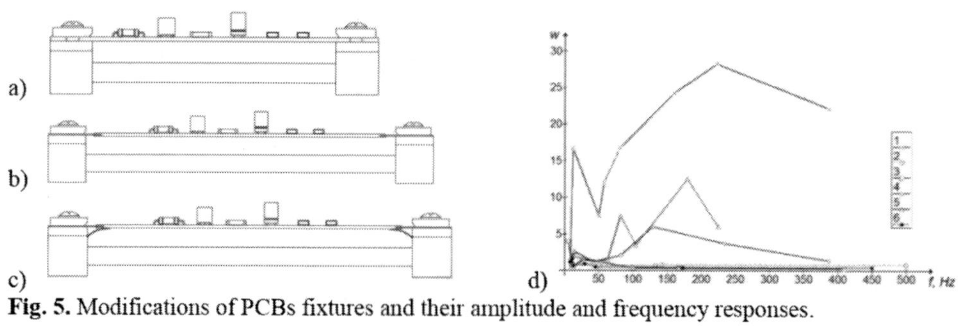


**Fig. 5.** Modifications of PCBs fixtures and their amplitude and frequency responses.

The dry friction damper was made from 0.3 mm thick bronze plate of 40 mm width and 15 mm length. The clamping force applied by the plate to PCB made 1.48 N. The value of this force was found with consideration of the plate deflection measured by the clock indicator and elasticity modulus.

Analysis of the complex application of the most effective methods, represented before, given as amplitude and frequency responses (fig. 6, a) and diagram of resonance values (fig. 6, b) for vibration isolating fabric tape with dry friction damper as a PCB fixture method and mentioned types of elastic and dissipative joints inside and outside of the electronic package enclosure, brought to the following statements:

1. using flexible installation of the case on the shaker governs increasing of vibration excitation of PCBs irrespective to types of mechanical joints inside the case. Herewith the highest degree of such excitation is observed for banding;
2. using bounding rail demonstrates the highest efficiency to reduce vibration when the case is firmly set on shaker as rigid installation. This is explained by the maximal level of structural damping provided by mutual movement of contacting surfaces introduced by dry friction damper;
3. increasing level of vibration excitation for flexible installation of the case on the shaker is explained by reduction of structural damping of vibration caused by insignificant mutual movement of contacting surfaces.

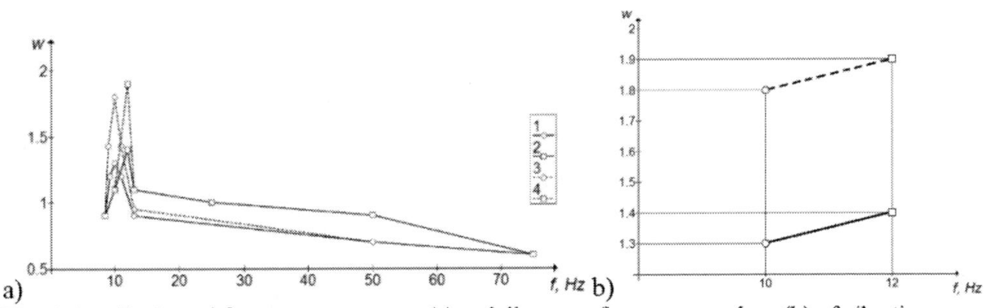


**Fig. 6.** Amplitude and frequency responses (a) and diagram of resonance values (b) of vibration transmission from the case supports to PCBs, fixed by using vibration isolating fabric tape with dry friction damper, in combination with using elastic and dissipative joints inside the case (∫ – banding; ○ – bounding rails) and outside the case (solid line - rigid installation; dash line - flexible installation).

## 5 Vibration isolation of the whole enclosure case

In order to secure the whole electronic package, including all components, subjected to external vibration another attempt was made by offering seismic enclosure case suspension. The suspension is introduced by insertion of rubber pneumatic chambers in between external and internal cases of enclosure, so their elastic joint is maintained by pressure in the chambers. In such kind of joint adjusting chamber pressure controls resonance frequency of the whole system and shifts it away from the resonance.

The experimental tests of the offered idea were conducted on the shaker for the following variants:

1. rigid installation of the case on the shaker;
2. foam rubber is inserted between PCBs;
3. the case is installed inside vibration isolation case, pneumatic chambers are under high pressure, foam rubber is inserted between PCBs;
4. the case is installed inside vibration isolation case, pneumatic chambers are under high pressure;
5. the case is installed inside vibration isolation case, lower pneumatic chambers are under half-pressure.

For all the variants the case was fixed on the shaker by rigid installation.

The test procedure was identical to the previous research. Vibration accelerations were read from two piezoelectric sensors: one attached to the center of PCB and another – to the shaker's platform. Amplitude and frequency response of vibration transmission shown in fig. 7 indicated the variant 3 as the most effective protection method against vibration.

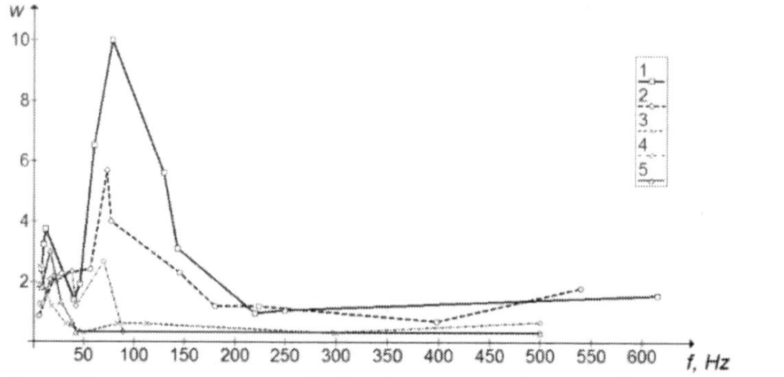


**Fig. 7.** Amplitude and frequency responses of vibration transmission from the case supports to $PCB_1$, for variants of vibration protection.

The advanced variant of vibration isolation was introduced by using dry friction damper (DFD) with suspension of the internal case by the soft springs inside the external enclosure (fig. 8, a) so that the natural frequency of the suspended case would make 3-8 Hz. Vibration at these frequencies is reduced by non-linear DFD, whose design is introduced by the following components: two plate springs 2, set in the bottom of external case 1, create wedge-shaped guides, which accommodate the rod 3 fixed to the internal case 4. The rod has two disks 5, made from hard rubber or fiberglass, attached to its surface.

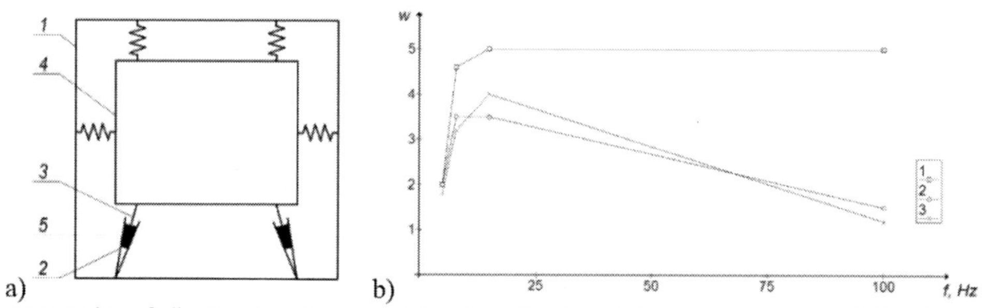


**Fig. 8.** Design of vibration isolation setup (a) and amplitude and frequency responses of vibration acceleration (b)

On the low frequency and high amplitude resonance of the external case, the wedge-shaped guides 2 moving along the rod 3 overcome friction force and dissipate vibration energy. The design of DFD illustrates that higher vibration amplitude increases movement of one case against another and hence increases the friction force because of growing normal pressure produced by the plate springs onto the rod. Quite to the contrary, on the high-frequency vibration, when amplitudes are small, DFD is not engaged at all.

The effectiveness of represented vibration isolation is demonstrated (fig. 8, b) by values of vibration acceleration measured on the external case (curve 1), internal case (curve 2) and printed circuit board (curve 3) correspondently during vibration tests.

## 6 Mathematical model of suspension on the fabric tape with dry friction damper

Analysis of experimental data has verified that the most effective and simple methods for protection against vibration and hereby providing strength and reliability of enclosed electronic packages subjected to external dynamic forces, like vibration, are: the installation or suspension on the fabric tape with dry friction damper for PCBs; and vibration isolation of the whole electronic package by using dry friction damper.

Suspension of PCBs on the fabric tape with dry friction damper gives the range of advantages: 1) PCB undergoes vibration in narrow frequency diapason – a few Hertz, depending on its weight. At higher frequencies outside of this diapason PCB remains still due to inertia and flexibility of the fabric tape – such structures are called as mechanical low-frequency filters; 2) low frequency vibration in the system proceeds without deformations of PCB, what secures through-hole and surface mount components; 3) easy access to components makes the assembly of such package maintainable unlike encapsulated packages sealed by compounds.

Since the damping in the fixture (fig. 9, a) is performed by friction of the friction plate 2 against the bearing pad 1, the magnitude of energy dissipation depends on friction force, vibration amplitude and other characteristics, which can be found from solution of differential equation that describes vibration of this system.

In mathematical model the fixture design is introduced by scheme that represents plate or beam oscillation on two elastic and damping supports (fig. 9, b, c). When stiffness of supports is considerably lower than that of PCB, PCB undergoes two resonances: 1st – by symmetrical form (fig. 9, b); 2nd – by skew-symmetric form of oscillation (fig. 9, c); further resonances are produced by deformations of PCB on the rigid supports.

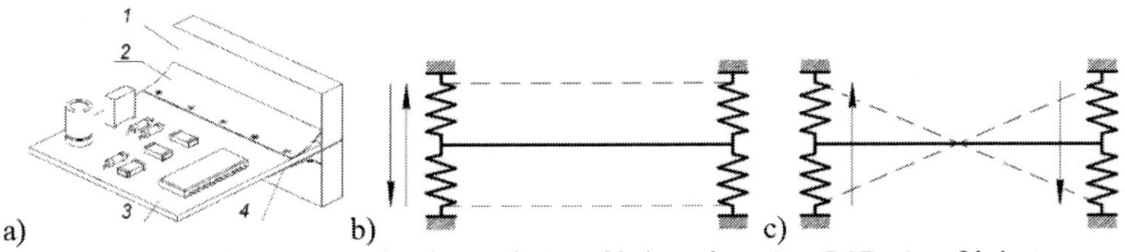


**Fig. 9.** Fixture design (a): 1 – bearing pad; 2 – friction plate; 3 – PCB; 4 – fabric tape, and PCB oscillation forms: b) symmetrical; c) skew-symmetric

Fig. 10 represents the 1st form of oscillation (fig 9, b), which is the most dangerous in condition of the offered fixture at low-frequency vibration. Mass *m* represents PCB 3 suspended by the spring 4 with stiffness *k*. The joint of the spring (case of enclosure) undergoes oscillation by the law $z = z_{\max} \sin\omega t$, which engages mass movement. Friction plates 2 are attached to mass. They are pressed to bearing pads by the force that produces specified deflection.

Symbol $x$ stands for PCB's movement, and $z$ – for the supports' movement, which is the forced oscillation. Stiffness of the spring $k$ is generalized parameter, which depends on tension (stiffness) of fabric tape, pressure force for friction plates, friction and other factors.

The mathematical model is brought to model of forced oscillations in single-mass system (fig. 10, a) [1,2] at presence of inelastic resistance forces produced by friction plates friction against surface of bearing pads:

$$R = \pm \frac{\alpha A^n}{\pi} \sqrt{1 - \left(\frac{x}{A}\right)^2} \tag{2}$$

where: $A$ – PCB's oscillation amplitude; $\alpha$, $n$ – constants, which depend on material.

Inelastic resistance force (2), indicated in the curve (fig. 10, b), represents deviation of the curve "force–movement" from linear Hook's law by hysteresis loop.

In accordance to [1,2] and considering (2) the forced oscillation of the single-mass system is described by equation:

$$-A\omega^2 \sin(\omega t - \gamma) + A\omega_c^2 \sin(\omega t - \gamma) + \frac{\alpha A^n}{\pi \cdot m} \cos(\omega t - \gamma) = \ddot{z} \sin \omega t \tag{3}$$

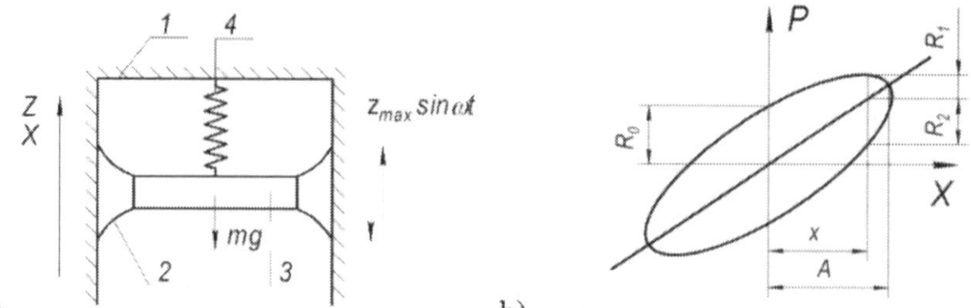


a) b)

**Fig. 10.** Oscillation model with dry friction damper (a) and force – movement diagram (b).

Assuming complexity of precise accounting effects for inelastic resistance force, which has nonlinear relationship with velocity $\dot{x}$, $R = b\dot{x}|\dot{x}|^{n-1}$, where $b$, $n$ – constants of the structure, it was substituted by energetically equivalent linear force $b_0\dot{x}$ using method of energy balance. Coefficient of equivalent inelastic resistance $b_0$ was found from condition of equal works for these forces in half-period of oscillation.

By the acceptable substitution of inelastic resistance force by the friction force $R = \pm F_{fr}$, when $n = 0$ (dry friction), amplitude is represented as:

$$A = \pm \frac{\ddot{z}}{\omega_0^2} \sqrt{\frac{1-\left(\frac{4F_{fr}}{\pi \cdot m \cdot \ddot{z}}\right)^2}{1-\left(\frac{\omega}{\omega_0}\right)^2}} \quad (4)$$

The first multiplier in the right part of equation represents static deviation, and the second one – dynamic factor, which has real values only in condition of $\frac{F_{\text{тр}}}{m \cdot \ddot{z}} < \frac{\pi}{4}$. Friction force is defined as:

$$F_{fr} = f_{fr} \cdot \delta \cdot E \frac{bh^3}{4l^3} \quad (5)$$

where: $f_{fr}$ – friction between surface of friction plate and bearing pads; $\delta$ – spring deflection.

Dimensions of the plate (spring) $l \times b \times h$ are specified by design considerations that comply with condition $F_{fr} < 0{,}785 m\ddot{z}$. Given width $b$, length $l$, material $E$ and stiffness $k = (2 \cdot \pi \cdot f)^2\, m$ of the plate its thickness is calculated as:

$$h = \sqrt[3]{\frac{4Cl^3}{Eb}} \quad (6)$$

By known friction the deflection of plate is found as:

$$\delta = \frac{F_{\text{тр}}}{2 f_{\text{тр}} N} \quad (7)$$

Obtained mathematical model provides optimization of oscillation system dynamic factor (4) with respect to geometric, elastic and dissipative characteristics of elastic and damping supports.

## 7 Conclusions

Vibration tests of different enclosure case modifications indicated that decreasing case stiffness increases vibration transmission to parts and components inside. Vibration transmission in case with walls removed is 2-4 times higher, and in case with walls from fiberglass – 1.5-2 times higher in comparison with regular case with aluminum walls.

High stiffness is required for building enclosure case even from damping materials in order to avoid additional resonances increasing vibration transmission inside the enclosure.

Building elastic and dissipative joints outside and inside of enclosure case demonstrated high effectiveness to reduce vibration transmission to printed circuit boards although its natural frequency does not change noticeably. The highest effectiveness was achieved by inserting bounding rail, which more than 5 times reduced vibration transmission.

Analysis of experimental data has verified that the most effective and simple methods for protection against vibration and hereby providing strength and reliability of enclosed electronic packages subjected to vibration, are: the installation or suspension on the fabric tape with dry friction damper for PCBs; and vibration isolation of the whole electronic package by using dry friction damper.

Obtained mathematical model provides optimization of oscillation system dynamic factor with respect to geometric, elastic and dissipative characteristics of elastic and damping supports.

## References

1. I. Kovtun, J. Boiko, S. Petrashchuk and T. Kałaczyński (2018, Jul.). Theory and practice of vibration analysis in electronic packages. *MATEC Web Conf.*, **182**, 02015 (2018)
2. J. Boiko, I. Kovtun and S. Petrashchuk, "Vibration transmission in electronic packages having structurally complex design," in *Proc. 2017 IEEE First Ukraine Conf. on Electrical and Computer Engineering (UKRCON)*, Kiev, pp. 514-517, (2017)
3. I. Kovtun, J. Boiko and S. Petrashchuk, "Identification of Natural Frequency and form of Oscillation for Electronic Packages Subjected to Vibration", in *Proc. 2018 IEEE 38th International Conf. on Electronics and Nanotechnology (ELNANO)*, Kiev, pp. 621 – 626, (2018)
4. H. Ardebili, J. Zhang and M. G.Pecht. *Encapsulation Technologies for Electronic Applications (Second Edition)*. United States, USA: William Andrew, Elsevier Inc., 508 pp., (2019)
5. E. J. R. Phua, M. Liu, B. Cho, Q. Liu, S. Amini, X, Hu, and C. LipGan. (2018, Mar.). Novel high temperature polymeric encapsulation material for extreme environment electronics packaging. *Materials & Design*, Vol. 141, pp.202-209, (2018)
6. I. S. Amiri, M.M. Ariannejad, D.Vigneswaran, C.S.Lim and P.Yupapin. (2018, Dec.). Performances and procedures modules in micro electro mechanical system packaging technologies. *Results in Physics*, Vol.11, pp. 306-314, (2018)
7. S. Akbari, A. Lövberg, P. Tegehall, K. Brinkfeldt and D. Andersson (2019, Feb.). Effect of PCB cracks on thermal cycling reliability of passive microelectronic components with single-grained solder joints. *Microelectronics Reliability*, 93, pp. 61-71, (2019)
8. V. Royzman, S. Petrashchuk, I. Kovtun, A. Lokoshchenko Calculation of mechanical stresses in adjoint system of electronic component and compound and strength assessment. *Journal of Vibroengineering*, Vol. 15 Issue 1, p. 65-71, (2013)
9. I. Kovtun, J. Boiko, S. Petrashchuk, G. Bauriene, K. Pilkauskas. Effects of the strain transmission from the main board to the installed electronic components. *MECHANIKA*, Vol. 22(6), pp. 489-494, (2016)
10. G.S. Pisarenko, V.A. Agarev. *Strength of materials*. (Kiev: Technika, 1967)
11. A. Troszyński, M. Liss, *Possibilities of the electrochemical drilling process's (ECM) vibroacoustic analysis,* Matec Woc, **182**, 1-8, (2018)

# Analysis of changes in the value of selected lubricant characteristics during use

*Bogdan* Landowski [1,*], *Monika* Baran[1]

[1]University of Science and Technology in Bydgoszcz, Faculty of Mechanical Engineering, Al. prof. S. Kaliskiego 7, 85-796 Bydgoszcz, Poland

**Abstract.** The problems connected with assessment of value changes of selected lubricants during the operation are addressed in this study. The lubricants which have been analyzed are engine oils used in internal combustion drive units of transport means. Selected results of tests of engine oils used for lubrication of passenger car drive units have been presented. The values of kinematic viscosity were determined for different temperatures. In order to evaluate the impact of the oil operation in real working conditions on the viscosity parameters, oil samples were collected from selected vehicles. The vehicles had been used in natural conditions providing transport services. Measurements of kinematic viscosity were performed by means of a rotary viscometer.

## 1 Introduction

The basic function of lubricants is to provide the surfaces of elements that make up tribological nodes protection from excessive wear. Application of lubricants affects the processes of wear by reducing their intensity.

Lubrication oils are commonly used for lubrication of machine elements. Thanks to the so called lubricating film that is formed on the element surfaces the friction forces are reduced and the wear process is limited [1].

The authors of this study address the problems connected with assessment of value changes of selected characteristics of lubricants during their operation. Engine oils for combustion drive units used in transport means were chosen for analysis.

Engine oils make up the costs involved in transport means operation [2-3]. In the case of large lubricating systems and short time between the oil changes, the costs are of particular significance. As regards public bus transport systems the costs are very high [3-4]. Hence, the models of transport means operation processes often include the costs of lubricants [3-8]. The economic factors such as the type of lubricant to be used or frequency of its change are not considered in this study.

The function of oil is also to prevent impurities from depositing on the engine elements. Sludges (low temperature sludge), sealing waxes (high temperature residue) carbon deposits (thermal disintegration products) should be dissolved and distributed within the oil by dispersing-washing agents contained in the oil. Thanks to this, they should be removed from the engine upon changing the oil. Another function of the oil is to protect an engineering object from corrosion. The oil should not have a destructive impact on the drive unit sealing [9-12]. Hence, it is necessary to provide organoleptic assessment of the lubricated elements such as, oil sump or oil sucking pipe which can be useful for evaluation of oils used in real conditions of a drive unit operation.

Monitoring rheological parameters of oil can facilitate assessment of the engine elements potential damage. Reduction of the oil viscosity value can indicate content of fuel or coolant in it.

* Corresponding author: bogdan.landowski@utp.edu.pl

The study presents an analysis of selected results of tests of engine oils used to lubricate drive units of passenger cars. The tests included selected engine oils with different viscosity classes according to SAE. Among others, their kinematic viscosity value was determined for different temperatures. Oil samples collected from selected vehicles were tested to provide assessment of the impact of oils used in real working conditions on viscosity parameters. The analyzed transport means were being used in real conditions and were providing transport services.

## 2 Research methodology

1000 $cm^3$ volume samples collected from drive units of the analyzed vehicles referred to as used oils have been tested. Samples of unused oil referred to as virgin oil have been tested as well.

**Table 1.** Data concerning used oils.

| No. | OIL | CAR | | | | | Oilmileage [km] |
|---|---|---|---|---|---|---|---|
| | SAE | Make and model | Production - year | Engine | Mileage | | |
| | | | | | Upon being poured into the engine [km] | After being poured down from the engine [km] | |
| 1. | 5W-30 | VW Tiguan | 2011 | 1.4 TSI 150KM | 125 000 | 139 520 | 14 520 |
| 2. | 5W-30 | Peugeot Partner | 2003 | 2.0 HDi 90KM | 382 670 | 403 279 | 20 609 |
| 3. | 5W-30 | Skoda Yeti | 2010 | 2.0 TDI, 170KM, chip tuning up 190KM | 93 458 | 109 585 | 16 127 |
| 4. | 5W-30 | Mazda 3 | 2005 | 1.6 CiTD 109KM | 258 750 | 269 695 | 10 945 |
| 5. | 0W-40 | VW Golf VI | 2011 | 1.4 TSI, 122KM | 66 000 | 73 000 | 6000 |
| 6. | 5W-40 | Opel Astra H | 2004 | 1.6 EcoTec 105KM | 37 115 | 52 375 | 15 260 |
| 7. | 10W-40 | Audi A4 | 1998 | 1.9 TDI 110KM | 257 420 | 269 670 | 12 250 |

The oils were divided into two groups groups. The first group included oils of the same viscosity class though coming from different drive units and samples of virgin oil. In the second group there were oil samples of the same make and type (with the same trade mark). The second group included post-use oils and unused oils (virgin) of the same summer (later winter) viscosity class.

To collect a sample, the used oil with temperature of about 90° C was poured down from the engine to a vessel. It was carefully mixed and then a sample was taken. Seven samples of used oil were subjected to tests in a drive unit and four samples of unused oils were tested as well.

Each measurement of the analyzed quantities was repeated five times.

Table 1 shows selected characteristics of the vehicles from which the tested engine oil samples were collected.

Kinetic viscosity measurement was performed by means of a rotary viscometer FUNGILAB. Viscosity measurement of each temperature was repeated five times, at 10 minute intervals, for each sample. The oil used for tests did not contain air bubbles and was stabilized.

The viscosity index was calculated on the basis of measurements according to the directives of PN-ISO 2909:2009 norm.

## 3 Selected tests results

The selected results of tests of the first oil group are presented in table 2 and in figures 1 and 2 [13-14].

Table 2 shows the values of the arithmetic mean of the kinematic viscosity measurement results and the determined values of viscosity index of oil denoted according to SAE 5W30 [13-14].

Figure 1 shows a comparison of kinematic viscosity of unused 5W-30 oil with used oils coming from different drive units in positive temperatures, whereas in figure 2 in negative temperatures [13-14]. Table 3 shows explanations of the oil sample code symbols whose test results are illustrated in figure 2.

In order to illustrate the considerations, figure 3 shows a photograph of a used contaminated oil from Peugeot Partner vehicle, whereas figure 4 shows paper chromatography of oil from Peugeot Partner car after 72 hours [13-14].

**Table 2.** Results of kinematic viscosity measurements and determined values of viscosity index - 5W30 oil.

| **Analyzed characteristic** | **Unused oil** | **VW Tiguan** | **Peugeot Partner** | **Skoda Yeti** | **Mazda 3** |
|---|---|---|---|---|---|
| Kinematic viscosity in temperature 21 °C | 145.52 $mm^2/s$ | 144.9 $mm^2/s$ | 210.24 $mm^2/s$ | 117.34 $mm^2/s$ | 140.94 $mm^2/s$ |
| Kinematic viscosity in temperature 40 °C | 65.42 $mm^2/s$ | 62.68 $mm^2/s$ | 88,38 $mm^2/s$ | 51,46 $mm^2/s$ | 56,02 $mm^2/s$ |
| Kinematic viscosity in temperature 100 °C | 11.24 $mm^2/s$ | 11.14 $mm^2/s$ | 15.6 $mm^2/s$ | 10.2 $mm^2/s$ | 11.02 $mm^2/s$ |
| Kinematic viscosity in negative temperature | 5055.98 $mm^2/s$ − 25 °C | 5158.16 $mm^2/s$ − 25 °C | 7590.62 $mm^2/s$ − 20 °C | 4988.22 $mm^2/s$ − 26 °C | 5585.92 $mm^2/s$ − 30 °C |
| Viscosity index in positive temperature | 166.03 | 172.29 | 188.46 | 190.98 | 193.34 |

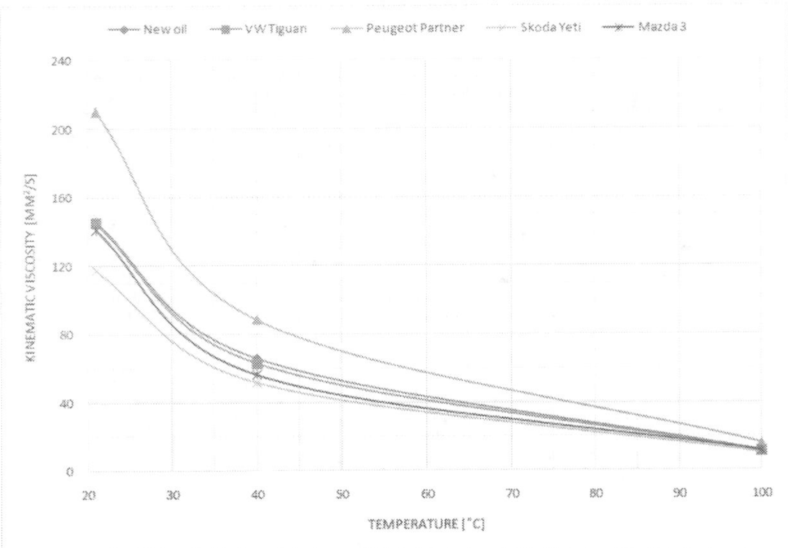


**Fig. 1.** Comparison of the values of kinematic viscosity of unused 5W-30 oil with used oils from different drive units in positive temperatures.

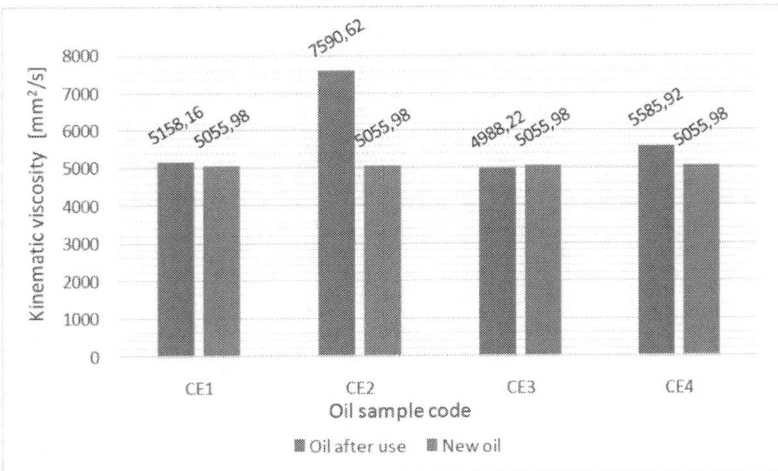


**Fig. 2.**Comparison of values of kinematic viscosity of unused 5W-30 oil with used oil from different drive units in negative temperatures.

**Table 3.** Code symbols of oil samples.

| Code | Explanation | Temperature of sample [°C] |
|---|---|---|
| CE1 | 5W-30 – VW Tiguan | - 25 |
| CE2 | 5W-30 – Peugeot Partner | - 20 |
| CE3 | 5W-30 – Skoda Yeti | - 26 |
| CE4 | 5W-30 – Mazda 3 | - 30 |
| - | 5W-30 – virgin oil (new oil) | - 25 |

**Fig. 3.** Contaminated oil from Peugeot Partner [14].

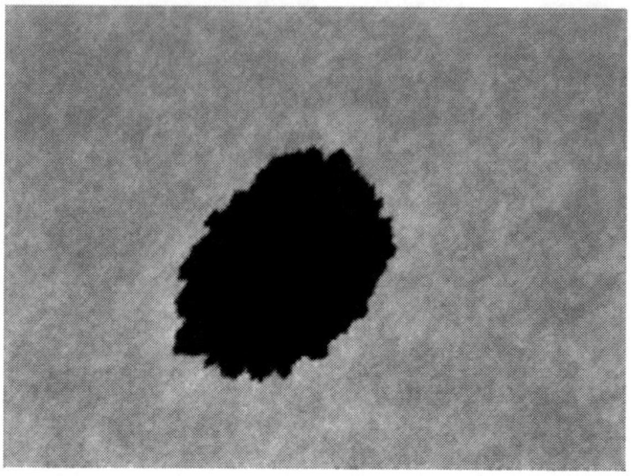

**Fig. 4.** Paper chromatography of oil from Peugeot Partner after 72 hours [14].

Selected results from tests of the second group of the analyzed oils are presented in table 4 and in figures 5 and 6.

An analysis of the diagram presented in figure 5 shows that the determined characteristics of oils are similar. Differences between viscosity values of oils decrease along with the temperature increase for all the analyzed temperatures.

**Table 4.** Mean values from kinematic viscosity test results and determined values of viscosity index.

| Tested characteristic | Result of measurement for 0W-40 oil | Result of measurement for 5W-40 oil | Result of measurement for 10W-40 oil |
|---|---|---|---|
| Kinematic viscosity in temperature 21 °C | 149.9 mm$^2$/s | 174.86 mm$^2$/s | 228.22 mm$^2$/s |
| Kinematic viscosity in temperature 40 °C | 66.12 mm$^2$/s | 75,38 mm$^2$/s | 94.98 mm$^2$/s |
| Kinematic viscosity in temperature 100 °C | 12.1 mm$^2$/s | 12.58 mm$^2$/s | 13.68 mm$^2$/s |
| Viscosity index in negative temperature | 3968.68 mm$^2$/s – 25 °C | 5015.78 mm$^2$/s – 25 °C | 6200.21 mm$^2$/s – 25 °C |
| Viscosity index in positive temperature | 166.03 | 166.7 | 145.65 |

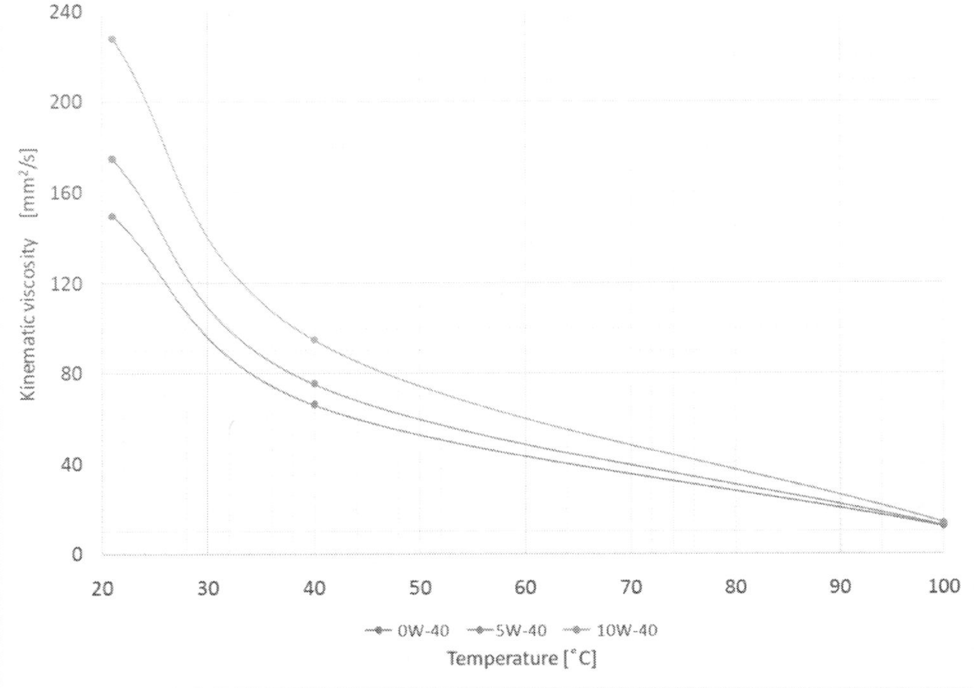


**Fig. 5.** Comparison of kinematic viscosity value for virgin oils from the analyzed group of oils.

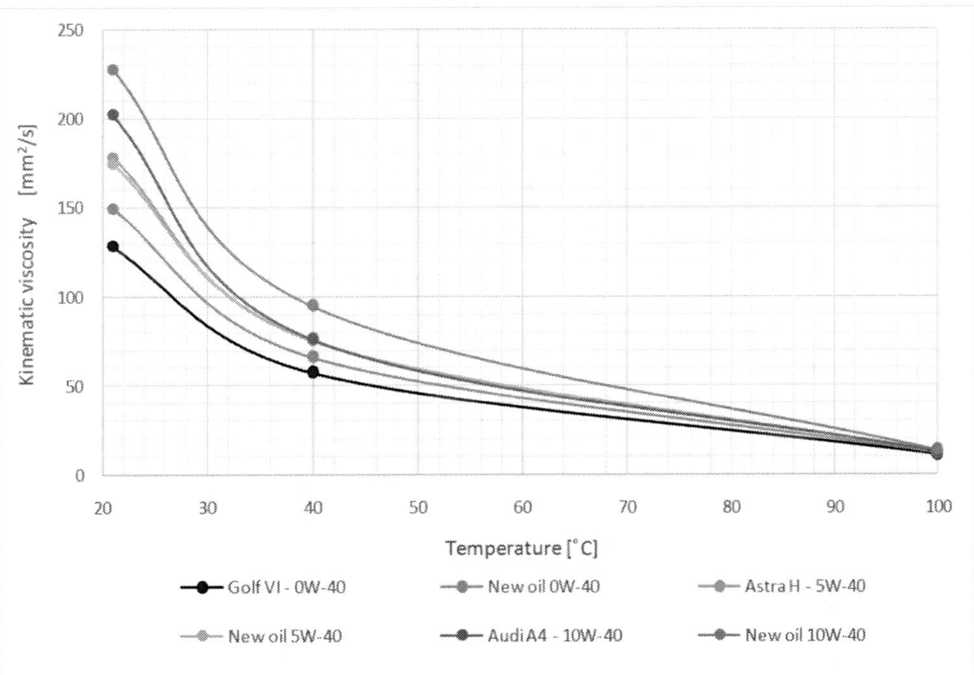


**Fig. 6.** Values of kinematic viscosity of oils with viscosity symbols according to SAE, respectively, 0W-40, 5W-40 and 10W-40 unused and used in drive units of the analyzed vehicles.

## 4 Conclusions

The test results presented in this study are supposed to provide assessment of the impact of engine oil application on value changes of selected characteristics of lubricants.

Basing on the result analysis it can be observed that kinetic viscosity value of the same type of oils can differ depending on the operation time and the drive unit they have been used in.

The highest values of kinematic viscosity for all tested temperatures were found for the oil from Peugeot Partner vehicle.

The dependence of the kinematic viscosity value on temperature has similar nonlinear characteristics for both used and unused oils. The differences between measurement results for virgin and used oils are lower in higher temperatures regardless of their viscosity class, and they increase along with the temperature drop for all tested temperatures.

An analysis of the test results allows to state that the highest increase in the oil kinematic viscosity of the used oil as compared to the virgin oil was found for 5W-30 oil from Peugeot Partner vehicle. It can be caused by impurities that are present in the engine in the form of sealing waxes and residues which were found by means of organoleptic assessment of the engine parts.

## References


1. W. Zwierzycki, Fuels, Oils, Motor opertional liquids. Oil Refinery 'Glimar" S.A. Editorial Office in Gorlice, Gorlice (1998)
2. B. Landowski, Ł. Muślewski, M. Pająk, O. Polishchuk, Method for initial assessment of unit costs of public city transport means operation, MATEC Web of Conferences 182, 01010 (2018), 17th International Conference Diagnostics of Machines and Vehicles, pp.1-7 (2018), https://doi.org/10.1051/matecconf/201818201010
3. B. Landowski, D. Perczyński, Ł. Muślewski, P. Kolber, Economic aspects of a city transport means purchase. Proceedings of 58th International Conference of Machine Design Departments - ICMD 2017, Publisher: Czech University of Life Sciences Prague, Czech Republic, pp. 194-199 (2017)
4. B. Landowski, Ł. Muślewski, Decision model of an operation and maintenance process of city buses, Proceedings of 58th International Conference of Machine Design Departments - ICMD 2017, Publisher: Czech University of Life Sciences Prague, Czech Republic, pp. 188-193 (2017)
5. B. Landowski, Numerical simulation of the process of a technical object state changes. Journal of KONBiN 45(2018), pp. 247-266 (2018)
6. B. Landowski, D. Perczyński, P. Kolber, Ł. Muślewski, An example of Markov model of technical objects maintenance process. Engineering Mechanics 2016, 22nd International Conference, may 9 – 12, 2016, Svratka, Czech Republic, Book of full texts, Institute of Thermomechanics Academy of Sciences of the Czech Republic, pp. 346-349 (2016)
7. B. Landowski, Ł. Muślewski, Numerical simulation of stochastic process as a model of technical object state changes. Engineering Mechanics 2018, 24nd International Conference, may 14 – 17, 2018, Svratka, Czech Republic, Book of full texts, Institute of Theoretical and Appiled Mechanics of the Czech Academy of Sciences, Prague, pp. 485-488 (2018)
8. B. Landowski, Numerical simulation of the process of a technical object state changes. Journal of KONBiN 44(2017), pp. 119-140 (2017)
9. A. Ambrozik, J. Jakóbiec, G. Wysopal, Basic tendencies in motor lubricant changes. Buses: technology, operation, transport systems, No. **12** (2011)
10. K. Hoag, B. Dondlinger, Engine Lubrication. Vienna (2015), DOI 10,1007/978-3-7091-1859-7_12

11. A. M. Ryniewicz, Ł. Bojko, T.Madej, Assessment of motor oil viscosity with the use of rotary viscometer, Scientific bulletins of Silesian University of Technology, Transport, Vol. **83**, No. 1904 (2014)
12. W. Urzędowska, Z. Stępień, Selectes issues concerning motor lubricant viscosity changes during operation. NAFTA-GAZ, December, Poland (2012)
13. B. Landowski, M. Baran, Analysis of viscosity of selected motor lubricants. University of Science and Technology materials, Department of Maintenance and Transport, Bydgoszcz, Poland (2019)
14. M. Baran, Analysis of selected characteristics of engine oils. Diploma thesis, University of Science and Technology, Bydgoszcz, Poland (2019)

# Analysis of selected results of engine oil tests

*Bogdan* Landowski [1,*], *Monika* Baran[1]

[1]University of Science and Technology in Bydgoszcz, Faculty of Mechanical Engineering, Al. prof. S. Kaliskiego 7, 85-796 Bydgoszcz, Poland

**Abstract.** The study presents selected results of viscosity tests performed for different temperatures of lubricating oil with viscosity marked as 5w30. Viscosity tests of new oil and oil right after being used have been compared. Lubricating oil used in a drive unit of a passenger car was tested. A vehicle in which oil had been changed irregularly was purposefully selected for the tests. Its mileage was over 15-20 thousand kilometers. Upon testing the vehicle mileage was above 265 thousand kilometers. The values of selected characteristics of the analyzed engine oils have been determined including: density, kinematic viscosity and viscosity index. FUNGILAB rotary viscometer was used for measurement of the oil kinematic viscosity.

## 1 Introduction

Most road transport means are equipped with internal combustion engines. Combustion engines are drive systems for passenger cars, trucks, coaches, and public transport buses. Engine oils are used to lubricate elements of both spark-ignition and self-ignition engines.

Engine oil is one of the most commonly used operational liquids in vehicles as without it the drive unit would be unable to work properly. Apart from its basic function, a lubricating oil has a number of other functions. Engine oil is also supposed to carry away the heat, seal and clean the engine, protect its metal parts from corrosion and reduce the destructive effect of the drive unit sealings. Monitoring of the oil state in a drive unit can be useful for assessment of its damage and defects [1-2].

One of the most important parameters of lubricating oil is its viscosity. Viscosity is a measure of internal friction between the oil particles moving relative to each other. In order to choose and compare engine oils its dynamic and kinematic viscosity is to be determined [3-4].

Excessive changes in the oil viscosity value can indicate damage of the drive unit and/or its defects. Significantly lower viscosity can indicate the presence of fuel or coolant in it. The content of a coolant can indicate, i.e. a damage to the gasket under the head, whereas the presence of fuel can mean, i.e. lack of tightness in piston rings [1].

Engine oils are part of transport means operational costs [5-6], and their manufacturing and utilization is an ecological issue. Therefore, people involved in transport undertake attempts to extend the vehicle mileage or a different measure of transport task accomplishment between successive oil changes.

In the case of large lubricating systems and a short time (mileage) between particular oil changes, the problem of engine oil related costs is of particular importance. This is a significant expenditure for public bus transport systems [6-7]. Hence, the models of transport means operation processes often include the costs of lubricating agents [6-10].

This study shows selected results of viscosity tests performed for different temperatures of the engine oil with viscosity 5w30. An attempt was undertaken to simulate the impact of the drive unit state on the engine oil viscosity characteristics. The results of viscosity tests

* Corresponding author: bogdan.landowski@utp.edu.pl

were compared for unused oil referred to as 'virgin' with oil right after use, referred to as 'used oil'.

## 2 Method of tests

The research object was the so called virgin oil, that is, oil that had not been used yet and a used oil coming from a drive unit of a passenger car with a given mileage and operation time.

The tested oil had been used in a Citroen Berlingo car. The vehicle selection criterion was irregular oil changes, after mileages longer than 15-20 thousand kilometers. At the time of tests the vehicle mileage was over 265 thousand kilometers. One of the objectives was to determine the impact of the engine state, including accumulation of sealing waxes and residue in the engine, on the oil viscosity characteristics.

An error indicator displayed in the vehicle indicating too low oil pressure. Moreover, the drive unit was characterized by excessive noise emission while operating. Virgin oil was introduced into the engine which after travelling 300 km was poured down. A test sample was taken from this oil. The analyzed vehicle was provided with new engine oil.

The samples were taken while the used oil was being replaced with new oil. The used oil whose temperature was nearly 90°C, was poured down from the engine to a clean vessel. Before the sample was taken it had been carefully mixed. The volume of the sample was 1000 $cm^3$. Unused oil was kept in a hermetically closed container to protect it from contact with air and contamination which could have a negative effect on the test results.

Glass beakers of 600 ml. volume were used to determine the oil density. An empty glass beaker was weighed to determine density of the tested sample and next, it was filled with the analyzed oil up to the volume of 600 $cm^3$ to be weighed again. This was repeated five times. The measurement results were rounded to 0.01g. On the basis of the measurements made, the density of oil was calculated.

A sample of 600 ml volume was poured into a vessel whose diameter was 97 cm. Viscosity measurement of each oil sample was repeated 5 times, every 10 minutes for each temperature. Kinematic viscosity test was performed with the use of a rotary viscometer FUNGILAB SMART series V210003. The tested liquid was stabilized and did not contain air bubbles [2, 11-12].

Moreover, assessment of the analyzed oils was performed by the method of chromatography. This involved putting a drop of used oil heated to the engine operation temperature on a blotting paper (120-130°C). This method is used for determination of presence of different impurities in the oil. An analysis involves observing the drop disintegrating into fractions. The colour of the stain nucleus made it possible to determine the amount of residue, the darker it was the more residue there was. A transparent/light yellow crown to be seen around the nucleus indicates fuel content in the tested oil. The diffusion ring width and the comminution level of carbon origin particles in this zone indicate the oil dispersing abilities. Light yellow or transparent border of the stain increases along with decreasing dispersing-washing abilities. The scheme of the stain is presented in figure 1 [2].

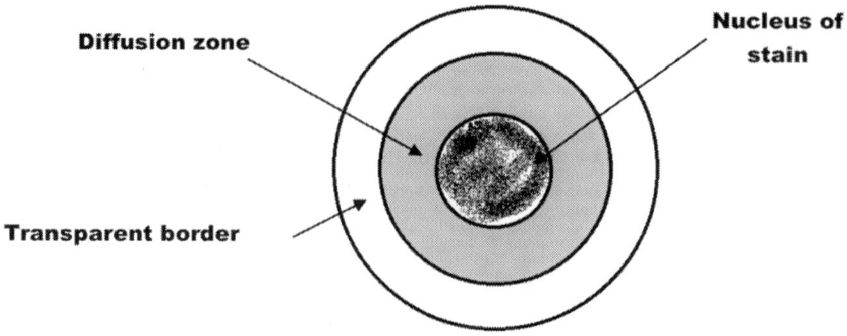


**Fig. 1.** A scheme of the oil stain on a blotting paper [2, 12].

## 3 Selected tests results

As part of the research, the values of characteristics of the analyzed engine oils including density, kinematic viscosity and viscosity index were determined. Selected results of the tests are presented in tables 1 to 5 and in figure 2 [11-12].

In table 1 there are results of the analyzed sample mass measurements which were later used to estimate density of the oils [11-12].

Table 2 shows the results of the tested oil kinematic viscosity measurements [11-12].

**Table 1.** Results of measurements of the analyzed sample mass.

| Measurement / Weighed element | $m_1$ [g] | $m_2$ [g] | $m_3$ [g] | $m_4$ [g] | $m_5$ [g] | Meanvalue [g] |
|---|---|---|---|---|---|---|
| Oil after use | | | | | | |
| Glass beak | 175.07 | 175.05 | 175.06 | 175.09 | 175.08 | 175.07 |
| Glass beak +600 ml oil | 686.79 | 686.8 | 686.79 | 686.84 | 686.83 | 686.81 |
| New oil | | | | | | |
| Glass beak | 179.87 | 179.86 | 179.87 | 179.86 | 179.86 | 179.864 |
| Glass beak +600 ml oil | 679.79 | 679.79 | 679.79 | 679.79 | 679.79 | 679.79 |

Table 3 shows the values of the calculated standard deviation and the coefficient of variability of viscosity measurement results [11-12].

Comparison of the results obtained from the tested unused and used oil samples is presented in table 4. Figure 2 illustrates the dependence of kinematic viscosity of the analyzed oils on temperature [11-12].

Based on the analysis of the oil kinematic viscosity dependence on temperatures (fig. 2) one can say that viscosity characteristic in a function of temperature is similar for both samples. The values of kinematic viscosity increase along with the temperature decrease.

**Table 2.** Results of kinetic viscosity measurement.

| Measurement / Temperature | $p_1$ [mm²/s] | $p_2$ [mm²/s] | $p_3$ [mm²/s] | $p_4$ [mm²/s] | $p_5$ [mm²/s] | Meanvalue [mm²/s] |
|---|---|---|---|---|---|---|
| Oil after use | | | | | | |

| | | | | | | |
|---|---|---|---|---|---|---|
| 21 °C | 169.5 | 171.1 | 170.1 | 172.2 | 172.8 | 17114 |
| 40 °C | 63.3 | 63.4 | 63.1 | 64.8 | 65 | 63.92 |
| 100 °C | 12.1 | 12.2 | 13 | 12.5 | 11.9 | 12.34 |
| -20 °C | 6259 | 6280.2 | 6290.2 | 6258.3 | 6250.1 | 6267.56 |
| New oil | | | | | | |
| 21 °C | 117.4 | 116.2 | 115.3 | 118.8 | 118,6 | 117.26 |
| 40 °C | 53.6 | 50 | 53.1 | 51.6 | 43 | 50.26 |
| 100 °C | 11 | 10.9 | 10.8 | 10.5 | 11 | 10.84 |
| -19 °C | 2143.6 | 2156.2 | 2187.6 | 2196.5 | 2140.9 | 2164.96 |

**Table 3.** Value of standard deviation and coefficient of variability of viscosity measurement results.

| **Temperature [°C]** | **Standard deviation [mm$^2$/s]** | **Variability coefficient [%]** |
|---|---|---|
| Oil after use | | |
| 21 °C | 1.383112432 | 0.808176015 |
| 40 °C | 0.903880523 | 1.414080918 |
| 100 °C | 0.427784993 | 3.466653101 |
| -20 °C | 16.85387196 | 0.268906432 |
| New oil | | |
| 21 °C | 1.512613632 | 1.289965574 |
| 40 °C | 4.295113503 | 8.545788903 |
| 100 °C | 0.207364414 | 1.912955844 |
| -19 °C | 25.58911878 | 1.181967278 |

The results of measurements for the sample of the used oil are significantly different from those provided for the virgin oil in the entire range. A distinct increase in the used oil kinematic viscosity as compared to the virgin oil was observed for negative temperatures. Although the post-use oil temperature was lower only by 1 °C than the temperature of the virgin oil, the value of the used oil kinematic viscosity was higher by as much as 4102.6 mm$^2$/s. The value of viscosity index for the used oil dropped as compared to the result obtained from a viscosity analysis of the unused oil.

**Table 4.** Presentation of the results obtained from tests of the oils.

| **Testedcharacteristic** | **Result of post-use oil measurement** | **Result of new oil measurement** |
|---|---|---|
| density | 0.8529 g/cm$^3$ | 0.8332 g/cm$^3$ |
| Kinematic viscosity in temperature 21 °C | 171.14 mm$^2$/s | 117.26 mm$^2$/s |
| Kinematic viscosity in temperature 40 °C | 63.92 mm$^2$/s | 50.26 mm$^2$/s |

| | | |
|---|---|---|
| Kinematic viscosity in temperature 100°C | 12.34 mm²/s | 10.84 mm²/s |
| Kinematic viscosity in negative temperature | 6267.56 mm²/s – 20°C | 2164.96 mm²/s – 19°C |
| Viscosity index | 194.96 | 214.02 |

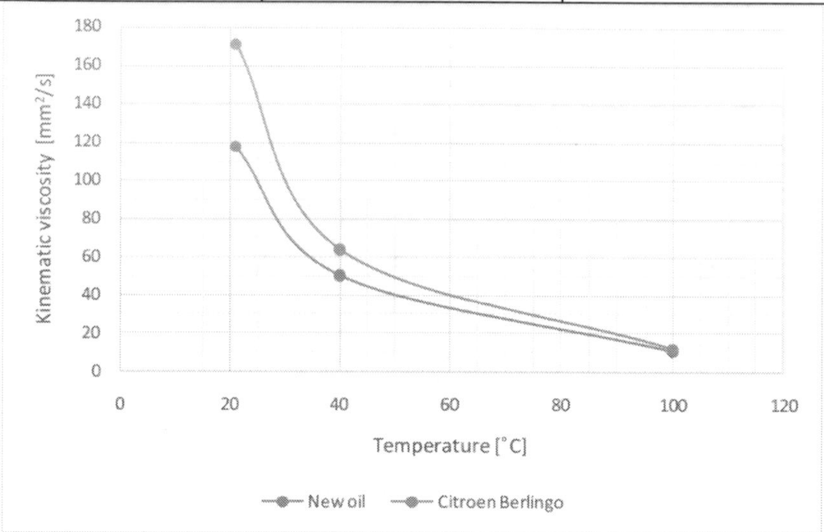


**Fig. 2.** Dependence between kinematic viscosity and the oil kinematic temperature.

No disturbing changes were observed during paper chromatography (fig. 3). The oil disintegrates into fractions in a proper way but it contains a lot of contaminants. The boundary of the diffusion and the stain nucleus is not visible. The blot test did not reveal the presence of water or fuel in the oil sample.

In order to illustrate the state of a Citroen Berlingo drive unit used for tests, figures 4 to 6 show photographs taken during the tests [11-12].

It seems that the tests results have provided the basis to formulate a conclusion of a universal character in the form of a recommendation. It appears to be advisable for vehicles with significant mileages to apply engine rinsing during a periodic oil change in order to remove the residues left on its elements.

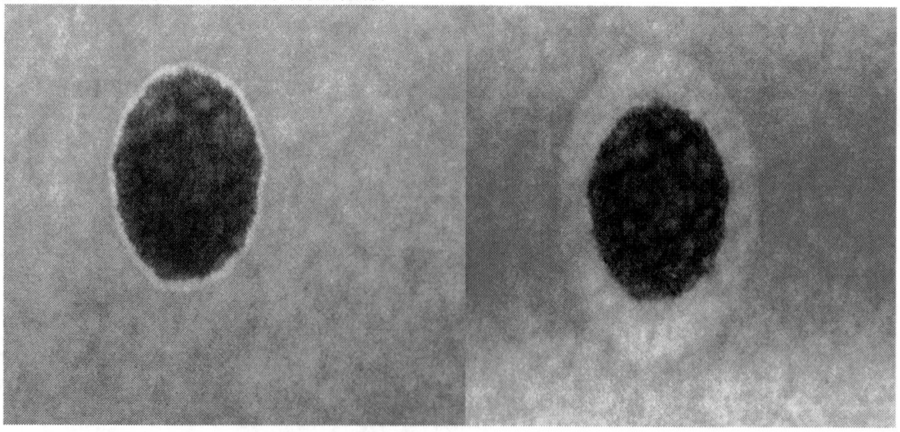

**Fig. 3.** A picture of chromatography on a paper after 24h and 72h [11-12].

**Fig. 4.** Filter of the oil suction pipe before and after cleaning [11-12].

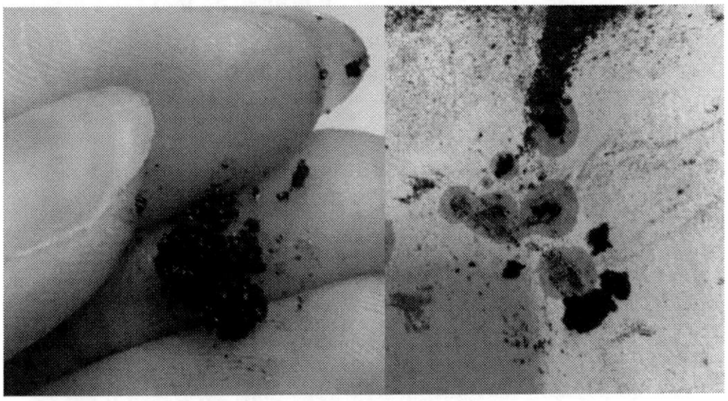

**Fig. 5.** Pollution of the oil sump bottom and filter of the sucking pipe [11-12].

**Fig. 6.** The engine bottom [11-12].

## 4 Conclusions

Comparison of the results of tests of used oils with virgin oils indicate that value changes in the oil rheological parameters can depend on the state of the engine which is indicated by the results of tests obtained for the oil used in a Citroen Berlingo car.

Low values of the calculated standard deviation and variability coefficient for the measurement results indicate high reproductivity of the obtained measurement results.

Paper chromatography (that is the blot or stain test), can be used, among others, to find out whether the tested lubricating oil contains water or fuel.

## References

1. W. Urzędowska, Z. Stępień, Selectes issues concerning motor lubricant viscosity changes during operation. NAFTA-GAZ, December, Poland (2012)
2. W. Zwierzycki, Fuels, Oils, Motor opertional liquids. Oil Refinery 'Glimar" S.A. Editorial Office in Gorlice, Gorlice (1998)
3. A. M. Ryniewicz, Ł. Bojko, T.Madej, Assessment of motor oil viscosity with the use of rotary viscometer, Scientific bulletins of Silesian University of Technology, Transport, Vol. **83**, No. 1904 (2014)
4. A. Ambrozik, J. Jakóbiec, G. Wysopal, Basic tendencies in motor lubricant changes. Buses: technology, operation, transport systems, No. **12** (2011)
5. B. Landowski, Ł. Muślewski, M. Pająk, O. Polishchuk, Method for initial assessment of unit costs of public city transport means operation, MATEC Web of Conferences 182, 01010 (2018), 17th International Conference Diagnostics of Machines and Vehicles, pp.1-7 (2018), https://doi.org/10.1051/matecconf/201818201010
6. B. Landowski, D. Perczyński, Ł. Muślewski, P. Kolber, Economic aspects of a city transport means purchase. Proceedings of 58th International Conference of Machine Design Departments - ICMD 2017, Publisher: Czech University of Life Sciences Prague, Czech Republic, pp. 194-199 (2017)
7. B. Landowski, Ł. Muślewski, Decision model of an operation and maintenance process of city buses, Proceedings of 58th International Conference of Machine Design Departments - ICMD 2017, Publisher: Czech University of Life Sciences Prague, Czech Republic, pp. 188-193 (2017)
8. B. Landowski, Numerical simulation of the process of a technical object state changes. Journal of KONBiN 45(2018), pp. 247-266 (2018)
9. B. Landowski, Ł. Muślewski, Numerical simulation of stochastic process as a model of technical object state changes. Engineering Mechanics 2018, 24nd International Conference, may 14 – 17, 2018, Svratka, Czech Republic, Book of full texts, Institute of Theoretical and Appiled Mechanics of the Czech Academy of Sciences, Prague, pp. 485-488 (2018)
10. B. Landowski, D. Perczyński, P. Kolber, Ł. Muślewski, An example of Markov model of technical objects maintenance process. Engineering Mechanics 2016, 22nd International Conference, may 9 – 12, 2016, Svratka, Czech Republic, Book of full texts, Institute of Thermomechanics Academy of Sciences of the Czech Republic, pp. 346-349 (2016)
11. B. Landowski, M. Baran, Analysis of viscosity of selected motor lubricants. University of Science and Technology materials, Department of Maintenance and Transport, Bydgoszcz, Poland (2019)
12. M. Baran, Analysis of selected characteristics of engine oils. Diploma thesis, University of Science and Technology, Bydgoszcz, Poland (2019)

# Analysis of vibrodiagnostics methods in the technical state study of designed multimedia mobile scenes

*Marcin* Łukasiewicz[1,*] *Michał* Liss[1], *Natalia* Dluhunovych[2]

[1]University of Science and Technology in Bydgoszcz, Faculty of Mechanical Engineering, Al. prof. S. Kaliskiego 7, 85-796 Bydgoszcz, Poland
[2]Khmelnytsky National University, Faculty of Programming Computer and Telecommunication System, 11, Instytuts'ka str., Khmelnytskyi, 29016, Ukraine

**Abstract.** The paper presents the possibilities of using vibroacoustic methods in the study of the technical condition of designed multimedia mobile scenes. In particular, the possibility of implementing modal analysis methods in modelling and diagnostic research process has been presented. The use of virtual methods enables diagnostic tests both at the design stage and at the stage of normal operation, whereas modal methods help to explain the nature of the work of the element under investigation.

## 1 Introduction

The process of designing new technical objects always poses a great challenge and adventure for the constructor. Through the implementation of a new design, the constructor can imprint their mark on the new technical object, showcasing their knowledge and skills. Newly created objects often change the future of humanity. A well-designed machine often sets new trends and introduces changes in everyday human life, often a few decades after its creation - examples include airplanes, ships etc. These and other complex technical objects often improve from year to year becoming more reliable and useful [1].

The multimedia mobile scenes can be classified into technical and constructional complex technical objects. These are modern mobile constructions enabling the realization of concerts or other artistic events in the open air. Mobile scenes are complicated constructions mounted on a car chassis, equipped with electrical and hydraulic systems. Event organisers of music concerts or other autistics events often allow mobile scenes to be installed in a designated place. The construction of mobile scenes requires the designer to carry out the appropriate design process including already at the design stage of diagnostic systems and procedures that ensure proper security of service and use of this facilities.

Modern machines are characterised by such features as: functionality, safety, reliability, mobility and exploitational susceptibility.The purpose of machine diagnostics is therefore to determine the technical condition of the object based on the generated diagnostic signals (symptoms) and compare them with the nominal values. An example of a mobile scene is shown in Figure 1.

* Corresponding author: mlukas@utp.edu.pl

**Fig. 1.** Multimedia mobile scene.

One of the most frequently used diagnostic methods encompass vibroacoustic methods, and among them, often used in the study of complex technical constructions, modal analysis methods. Due to the complexity of the modal analysis methods, dedicated IT systems have been created enabling engineers to create a virtual model of any object. Then, if necessary, the engineering team can carry out research on both a virtual and real object. It creates the possibility of introducing diagnostic elements to the object at the stage of its construction, modifying it and making changes to the object without incurring financial expenses for creating costly prototypes.

## 2 Basis of vibrodiagnostics

The increasing complexity of modern technical facilities and the growing safety requirements force constructors and users of these facilities to supervise their current technical condition. Technical condition of designed object should be verified by diagnostics means and procedures. The basic problems of machine diagnostics include the following:

- acquisition and processing of diagnostic information,
- building models and diagnostic relations,
- diagnostic inference and limit values,
- classification of machine states,
- predicting time of the next diagnosis,
- imaging decision-making information [2,3,4,5,6].

The vibrodiagnostics is one of the machine technical condition description methods - organised set of methods and means to the technical state estimation of technical systems, with utilization of vibration processes or the noise signal [7].

Vibroacoustics is a discipline of knowledge focussing on all vibration processes, acoustic and pulsating processes setting in technique and machines. The vibroacoustics processes are not only harmful but they are also positive phenomenon. If applied suitably, they can be a good carrier of the energy which can be used for the realization of various technological processes [7].

The vibroacoustics processes are dynamic phenomena set in a technical device with frequency range from zero to tens thousands of hertz. Vibrations are a rich source of information about the threats of the machine's condition, which are comparatively easy to record, processing and identification. We can introduce vibration processes in the following aspects [7]:

- the generation of strengths variables in time, acting on the structure and surrounding environment,
- propagation and transformation of energy in structures,
- the sound radiation through the elements of environment and mechanical structures.

The investigations of vibroacoustics processes in many cases are very complicated, in particular when vibration processes step out in real physical arrangements. During analysis of vibroacoustics process we focus on the following aspects:

- the temporary and spatial schedule of energy coming from the source,
- the arrangement response and transfer by the propagating media,
- the correlation between sources [7].

Obtaining the right technical construction response during vibroacoustic tests is possible with the use of appropriate measuring equipment, specialised software and appropriate conditions during research. The individual stages of the processes involved during investigations are shown in Figure 2.

After selection of the appropriate measurement method is completed, the scientist conducting experimental research often faces a dilemma. Specifically, he or she needs to determine - in which place of the technical object there should be points of signal reception carrying information about the object's condition. Engineers know that each object, depending on the structure, has certain places considered appropriate for signal reception. However, there is always a high probability of an erroneous decision.

Geometric modelling, implemented in modern engineering software, may be very helpful. After entering the parameters of the technical object, the modelling process generate a grid of potential points where measurements of the examined physical quantity should have been made. Consideration should be given to significant technical relationship during tests. Specifically, the higher the degree of tested object complexity, the greater density of the network's measurement points should be installed in the object. Even computer programs are utilised, the number of points generated is usually too high, which entails a longer global measurement process. In each experiment, we strive to obtain an optimal geometric model which is usually based on the subjective feelings of the experimenter and his o her professional experience.

The assessment of the technical object condition is estimated on the basis of diagnostic tests, consisting of measurements and analyses of the results obtained during tests. In order to carry out diagnostic tests, models of diagnostic objects are adopted, based on which diagnostic algorithms are generated.

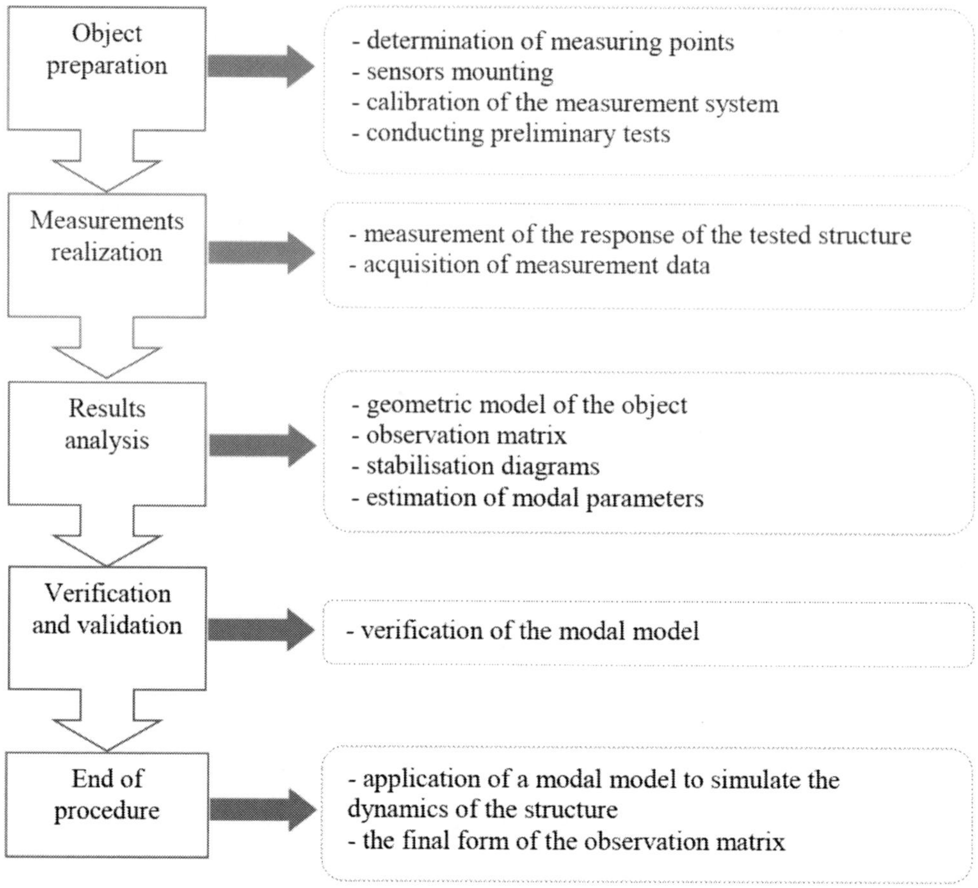


**Fig. 2.** Stages of proceeding in the research process.

## 3 Basis of modal analysis

Application of the modal analysis as one of the vibroacoustics tools constitutes a new approach to the technical analysis of the complex technical object such as multimedia mobile scenes. Modal analysis is the process of determining modal parameters of a structure for all modes in the frequency range of interest. Modal parameters are: modal frequency, modal damping and mode shapes. The modal analysis methods are divided into [10,11,12]:

- theoretical modal analysis - which requires its own solution to the problem for the adopted structural model of the object. Determined sets of own frequencies, damping coefficients for own frequencies and forms of natural vibrations allow simulations of the behavior of the structure at any extortion, selection of controls, modification of structures and others. It is used in the design process when it is not possible to conduct research on a similar real object,
- experimental modal analysis- controlled experiment requiring identification, during which forces the movement of an object (e.g. vibration) and measures the force and response in many measurement points distributed on the test object.

  Conducting experimental modal analysis requires measuring the vibrations of a structure excluded from motion at many of its points - at the same time, the force of vibration must be measured. Estimation of the parameters of the modal model consists in approximation of the measured characteristics of the structure by means of a function for which the variables are parameters of the modal model,

- operational modal analysis - based on operational experiment in which measurements are made only in response to the number of measurement points, while the movement of the subject is caused by the actual operating motion. Operational modal is the name for the technique of modal analysis performed on operational data - cases where we do not stimulate the structure artificially but just allow the natural operating loads to excite the structure.

Modal analysis of mechanical systems is therefore a method of testing dynamic properties of structures based on vibroacoustic signals. The idea behind this method is observation of the changes in parameters of the modal model that arise as a result of deregulation, normal usage, damage or failure, during current monitoring of the object. In this method, a modal model is created in the form of a set of own vibrations, mode shapes and damping coefficients, for the object without damage, as a pattern. Then, during operation, the modal model is identified and its correlation with the model for the undamaged object is examined. In the case where such a correlation occurs, it can be concluded that the object is in the good technical state. In the absence of correlation, the object is not working properly – for example due to damage [12,13,14].

Owing to the complexity of the structure characterised by the multimedia mobile scene, the theoretical modal analysis should be applied at the design stage. In case of building the prototype estimated modal models should be verified in real conditions with operational modal analysis tests carried out on the real object, during its normal operation. These tests will allow to verify the correctness of the adopted design assumptions and to make possible constructional changes for the designed mobile scene.

### 3.1 Operational modal analysis – method theory

In order to verify the real behavior of the multimedia mobile scene in the conditions of real operation, it is necessary to condct modal tests of the operational modal analysis. There are a number of methods used. However, the theoretical basis of operational modal analysis in this work is based on the Least Squares Complex Exponential (LSCE) method.

In LSCE method determination of the modal model parameters is followed by correlation function which is approximated by the sum of exponentially decaying harmonic functions. This method, applied to impulse response of system is a well known method in modal analysis yielding global estimators of system poles – the root of the transfer function denominator. It can be proved that the cross correlation function can be used to identify modal system parameters in identical way as does the impulse response of the system. The dynamic equation of the system motion can be expressed by formula [8,9]:

$$M\ddot{x} + C\dot{x} + Kx = F(t) \tag{1}$$

where: M,C,K – mass, damping and stiffness matrices, $\ddot{x}, \dot{x}, x$ – acceleration, velocity and displacement vectors, F(t) – vector of exciting forces.

The next step is transformation of formula (1) to principal coordinates applying the transformation expressed by the formula [8,9]:

$$x(t) = \Psi q(t) = \sum_{r=1}^{n} \Psi_r q_r(t) \tag{2}$$

where: $\Psi$ – matrix of modal vectors, the columns of which are eigenvectors corresponding to the given free vibration frequency, qr – principal (modal) coordinate, n – number of vibration forms included in the model of vibration forms.

Assuming that damping is small and proportional, on substituting relation (2) into formula (1) and multiplying by $\Psi$T, de-coupled equation set is obtained in the form [8,9]:

$$\ddot{q}_r(t) + 2\xi_r \omega_{nr} \dot{q}(t) + \omega_{nr}^2 q_r(t) = \frac{1}{m_r} \Psi_r^T f(t) \tag{3}$$

where: $\omega_{nr}$ – is the r-th free vibration frequency, $\xi_r$ – is the modal damping coefficient for the r-th vibration form, $m_r$ – is the modal mass.

Cross correlation function for two response signals at point i and j, resulting from an excitation applied at point k in the form of white noise has the following form [8,9]:

$$R_{ijk}(T) = E_o[x_{ik}(t+T)x_{ik}(t)] \tag{4}$$

where: $E_o$ – denotes the expected value operator.

If we know relation between cross correlation function, having the form of a sum of exponentially decaying harmonic functions and impulse transition function for direct application to do modal analysis tests, correlation function can be transformed to the form [8,9]:

$$R_{ij}(T) = \sum_{r=1}^{n} \frac{\psi_{ir} g_{jr}}{m_r \omega_{rd}} exp(-\xi_r \omega_{nr} T)\, sin(\omega_{rd} T + \vartheta_r) \tag{5}$$

where: $\vartheta_r$ – the New phase angle, $G_{jr}$ – constant.

### 3.2 Operational modal analysis – sample test results

Initial research was carried out at the University of Science and Technology in Bydgoszcz. The LMS Scadas Recorder with the attached LMS Test.Xpress software was used for the tests. The identification of modal parameters in the operational modal analysis is based only on the measurement of the system response. Identification methods are divided into those implemented in the time domain and those implemented in the frequency domain [10,11,12]. The apparatus used in the research makes it possible to use the following methods to identify modal parameters:

1) implemented in the field of time domain (module Op. Time MDOF):
- BR (Balanced Realisation), parameter estimation using the autoregression function of response signals in a stochastic subspace,
- LSFD (Least Squares Frequency Domain) – mode shapes estimation - parameter estimation using the non-linear least-squares method in the frequency domain,

2) implemented in the field of frequency domain (module Op. PolyMAX):
- pLSCFD (Polyreference Least Squares Complex Frequency Domain) – method implemented on the basis of operational transmittance,
- LSFD (Least Squares Frequency Domain) - mode shapes estimation - parameter estimation using the non-linear least-squares method in the frequency domain.

Estimation of modal parameters is one of the most important stages of operational modal analysis. The number of available methods for estimating modal parameters is enormous, but this does not mean that all of them will be perfectly suited to the research being carried out. Each of the applied methods has its own advantages and disadvantages, which in the case of the above mentioned are presented in Table 1.

**Table 1.** Advantages and disadvantages of estimating modal parameters[10,11].

| | |
|---|---|
| **Estimation in time domain** | |
| Advantages | more effective for noisy data |
| | additional errors resulting from spectrum leakages are avoided, etc. |
| | more effective in wide frequency ranges |
| Disadvantages | difficult analysis of systems with high damping |
| **Estimation in frequency domain** | |
| Advantages | a simpler way of data averaging |
| | more effective in the case of a large effect of the vibration form outside the frequency range |
| | much easier to determine the poles of the system in systems with high damping |
| Disadvantages | not very effective for wide frequency ranges |

The basis for modal analysis is the proper creation of a modal model of the real mechanical construction. Such a model is created during the course of modal analysis and is understood as a ranked set of natural frequencies, values of damping coefficients and the order of the modal model. In order to have any grounds for the assessment of the technical condition, the obtained modal model should be compared with the model created in the case of a machine in good working condition. Modal model duting test is obtainied on the basis of stabilisation diagram. Examples of stabilisation diagrams obtained for the same object using time and frequency estimation methods are shown in Figures 3 and 4. In the following stabilization diagrams, individual symbols are visible, which means: s - stable pole, o - unstable pole, f - constant frequency pole, v - constant frequency pole and modal vector, d - constant damping pole [8,9,13-15].

The modal parameters determined from the stabilisation diagram constitute a description of the modal model and are basic comparative parameters of a given technical object in the state of proper working condition. In the case of damage to a technical object, completely or slightly different modal parameters describing the modal model are obtained.

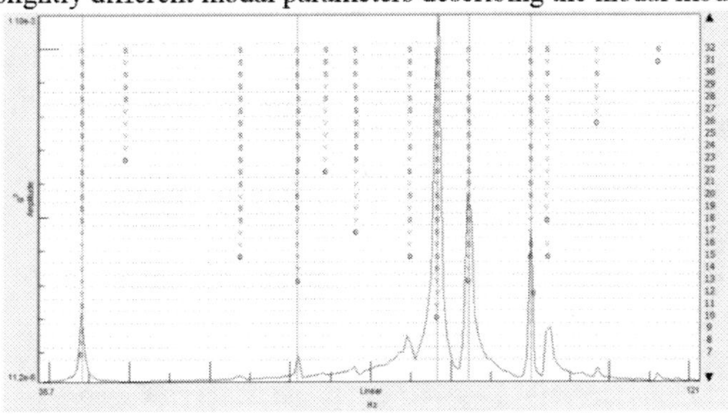

**Fig. 3.** Stabilisation diagram determined by the PolyMAX method.

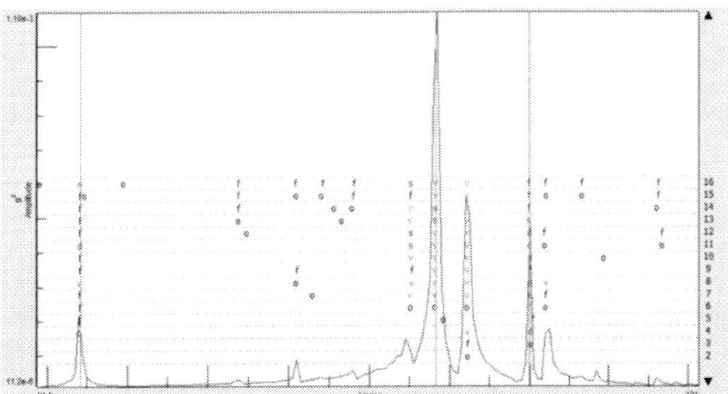

**Fig. 4.** Stabilisation diagram determined by the Time MDOF method.

An additional benefit to this process is the visualisation of individual forms of vibrations on the geometrical model of the diagnosed mechanical structure. The basis for identification issues is the appropriate verification of built models based on the collected experimental data. The verification process of the modal model consists of:

- visual checking of the model quality,
- study of the properties of coefficients describing the relationship between the object of study and the model,

- analysis of the suitability of the modal model, in the application for which the model was identified.

Based on the obtained modal patramerers data from two different states of the technical object, a diagnostic inference process can be carried out regarding the change in the condition of the machine's fitness.

## 4 Conclusions

At the moment of creating and designing new objects, diagnostic problems must have their application already at the construction stage and the knowledge of the constructor should be supported by the experiments of the diagnostic teams. The issue of diagnosing the condition of technical objects necessitates the diagnostitian possess an indepth diagnostics knowledge and to be up-to-date with a number of methods and diagnostic measures.

The diagnostic methods presented in this paper, based on the methods of vibrodiagnostics, appear to be viable and justified for the implementation of newly designed mobile scenes. Particularly noteworthy are the methods of modal analysis, which are discussed in the paper as an example of these modern tools supporting the work of the constructor. Identification of potential technicals problems, improvement of the quality of designed mobile scene construction based on modern software tools will eliminate a number of problems at the design stage. This will shorten the time to complete the scene and to ensure a high level of safety of the mobile scene.

This paper has been achieved under the research project "Hybrid multimedia mobile scenes are a chance for decisive innovation" No. POIR.04.01.04-00-0045/17-00.

## References

1. M. Dietrich, *Machine construction basics (Podstawy konstrukcji maszyn),* WNT, Warszawa, (2008)
2. C. Cempel, *Vibroacoustic diagnostics of machines (Diagnostyka wibroakustyczna maszyn).* PWN, Warszawa, (1989)
3. B. Żółtowski, M. Łukasiewicz, T. Kałaczyński, *Information technology in machine condition studies (Techniki informatyczne w badaniach stanu maszyn),* Wydawnictwa Uczelniane Uniwersytetu Technologiczno - Przyrodniczego w Bydgoszczy, (2012)
4. B. Żółtowski, M. Łukasiewicz, *Vibration diagnostics of machines (Diagnostyka drganiowa maszyn),* Wydawnictwo Naukowe Instytutu Technologii Eksploatacji, Bydgoszcz – Radom, (2012)
5. B. Żółtowski, C. Cempel, *Engineering of diagnostics machines* (in Polish), Polish Society of Technical Diagnostics, Warsaw, Bydgoszcz, Radom, (2004)
6. B. Żółtowski, T. Kałaczyński, *Machine diagnostics. Lecture and exercises (Diagnostyka maszyn. Wykład i ćwiczenia),* Wydawnictwo Uczelniane Uniwersytetu Technologiczno – Przyrodniczego, Bydgoszcz, (2013),
7. B. Żółtowski, M. Łukasiewicz, *Vibration assessment of the consumption of a technical object,* Vibroengineering PROCEDIA, **6**, p.109-114, (2015)
8. T. Uhl, W. Lisowski, *Exploitation modal analysis and its application (Eksploatacyjna analiza modalna i jej zastosowanie),* AGH Kraków, (1999),
9. T. Uhl, P. Kurowski, *Vioma - user's manual (Vioma – instrukcja użytkownika),* AGH Kraków, (2002)
10. W Heylen., S Lammens., P Sas, *Modal Analysis Theory and Testing,* Katholieke Universiteit Leuven, Heverlee, (2007)
11. T. Uhl, *Computer-aided identification of mechanical structure models (Komputerowo wspomagana identyfikacja modeli konstrukcji mechanicznych),* Wydawnictwo Naukowo – Techniczne, Warszawa, (1997)

12. M. Łukasiewicz, T. Kałaczyński, J. Musiał, J.Shalapko, *Diagnostics of buggy vehicle transmission gearbox technical state based on modal vibrations*, Journal of Vibroengineering, **16**, Issue **6**, Pages (2624-3168), (2014)
13. M. Pająk, Ł. Muślewski, B. Landowski, A. Grządziela, *Fuzzy identification of the reliability state of the mine detecting ship propulsion system*. Polish Maritime Research**, 1(101)** (2019)
14. J. Musiał, Ł. Muslewski, M. Pająk, A. Grządziela, *Analysis of vibration time histories in the time domain for propulsion systems of minesweepers*, Journal of Vibroengineering, **17, 3**, (2015)
15. M. Matuszewski, T. Mikołajczyk, D.Yu. Pimenov, M. Styp-Rekowski M, *Influence of structure isotropy of machined surface on the wear process,* International Journal of Advanced Manufacturing Technology, **88**, (2017)

# Impact of changes in the fuel injection controller settings on the power and torque performance depending on the engine speed

*Marietta* Markiewicz[1,*] and *Natalia* Dluhunovych[2]

[1]UTP University of Science and Technology, Faculty of Mechanical Engineering, Al. prof. S. Kaliskiego 7,- 85-796 Bydgoszcz
[2]Khmelnytsky National University, Faculty of Programming Computer and Telecommunication System, 11, Instytuts'ka str., Khmelnytskyi, 29016, Ukraine

**Abstract.** The internal combustion engine is the powerhouse of motor vehicles. The basic fuel used for self-ignition engine powering is diesel oil. The efficiency of an internal combustion engine depends on many factors and is defined by such parameters as power output, torque, and the engine rotational speed. The engine performance parameters can be improved by modification of the manufacturer's computer software that controls the engine operation. The study presents results of tests of a drive unit involving adjusting the fuel injection controller settings. The modifications involved increasing the fuel dose and the air charge pressure. Tests were carried out on a load bearing chassis dynamometer. The research object was a self-ignition engine with 109 KM power and Common Rail injection system. Power and torque measurements were carried out for the engine rotational speed range up to 4000 rev/min. It was found that modifications of the manufacturer's engine computer software contribute to power and torque increase for all the rotational speeds.

## 1 Introduction

The first self-ignition engine was constructed by Rudolf Diesel in 1893. It was powered by peanut oil. Not sooner than after development of petroleum refining technology diesel oil was found to be a much more universal fuel when used in self-ignition engines [1,2]. The twentieth century was a breakthrough in petroleum extraction and refining which made it the fastest developing branch of industry [3,4]. Deposits of sticky crude oil are sought by geologists and obtained through wells in a 1-kilometer deep soil, for which mining statistics are needed [5]. Advantages of fuels designed for these engines are their good lubricating properties, lower viscosity and high octane number which indicates the engine ability to self-ignite. According to PN-EN590+A1:2017-6 norm [6] diesel fuel should have specific parameters which are presented in table 1.

**Table 1.**Diesel oil parameters.

| | |
|---|---|
| Cetane number | 50 |
| Density (w temp. 15°C) | 820-845 kg/m$^3$ |
| Content of aromatic carbohydrates | ≥11% |
| Sulfur content | ≥10 mg/kg |
| Ignition temperature | 55°C |
| Water content | ≥200 mg/kg |
| Viscosity (in temp. 40°C) | 2-4,5 mm$^2$/s |

* Corresponding author: marmar000@utp.edu.pl

Although the structure of a modern drive unit is significantly different from its prototype constructed in the XIX century, the basic function of the engine, that is, unaided fuel self-ignition has not changed so far [7]. The structure of a modern self-ignition engine enables modification of the manufacturer 's computer software to improve the engine performance. There is a variety of methods to be used for the engine performance parameter improvement [8]. The most common methods involve connecting an additional device to the engine beam, the so called "power box" [9]. Another method is modification of the manufacturer's engine computer software [10, 11]. This method involves reading the engine characteristics and uploading a new software for the fuel injection controller setting. The software can be read from the vehicle computer through a diagnostic connector OBDII, directly from the processor through BDB connector by removing the memory and reading in the programmer [9, 10]. The manufacturer's settings are most often modified for drive units equipped with turbocharging, as they contain an electronic control unit [12].

This study aims at checking the impact of the engine computer software modifications on the torque increase depending on the drive unit rotational speed.

## 2 Materials and methods

The research object was a self-ignition engine mounted in a vehicle. The drive unit was selected due to its being frequently used in car transport, in vehicles with admissible gross vehicle weight up to 3.5 t. The engine was characterized by direct fuel injection, controlled by means of Common Rail system with electromagnetic injectors. Technical data of the drive unit are presented in table 2.

**Table 2.** Technical data of a drive unit.

| | |
|---|---|
| Engine cubic capacity | 1560 $cm^3$ |
| Injection type | direct |
| Engine power | 109 KM |
| Maximal torque | 240 Nm |
| Number of cylinders | 4 |
| Cylinder diameter | 73 mm |
| Piston stroke | 88,3 mm |
| Number of valves | 16 |
| Charging | Turbo compressor |

The engine which was tested was serviceable with a drive unit used in the conditions of a city traffic. Prior to starting the measurements the drive unit was heated up to the temperature of the coolant which was about 85°C. All the sensors and probes were connected to the vehicle (temperature sensor, pressure sensor, exhaust emission and particulate measurement probes). The structure of the drive unit used in the tests is characteristic of solutions applied in high power engines. Thanks to this, it is possible to adapt the obtained results to a wider group of engines.

The experiment was performed to determine the engine performance parameters. The tested object was placed in a chassis dynamometer equipped with an eddy current brake of Dynotech type simulating the road conditions. During the experiment the fuel injection settings were changed to acquire necessary information. Chassis dynamometer enables acquisition of data indispensable for determination of the drive unit performance characteristics such as power and torque [13, 14]. The device was used to measure torque on the crankshaft and calculate the engine power on the basis of the torque value and the crankshaft rotational speed [15]. The measurement was affected by the air moisture, atmospheric pressure changes and the air temperature whose values were normalized. The results provided by the measurements were calculated into values corrected according to the applicable norm (25°C and 1000 hPa). A test measurement was performed to make it possible for the dynamometer to automatically determine the location of gear box location of the gear box and the final drive for the gear which would be used for performance of the proper measurement. The chassis dynamometer used in the tests consists of a suspension system with the braking unit, the engine cooling system and a control-measurement system. Figure1 shows a vehicle which was mounted on the chassis dynamometer. Specification of the chassis dynamometer used for the tests is presented in table 3.

**Table 3.** Specification of chassis dynamometer.

| | |
|---|---|
| maximal speed | 288 km/h |
| Maximal torque on the wheels | 1300 Nm per one axle |
| Maximal power | 900 HP per one axle |
| Maximal torque on the engine | 1000 Nm per one axle |
| brake powering | 230 V/16A |
| mass | 3000 kg |
| Length/width/height | 3800/4500/500 mm |
| maximal/minimal | 2100/700 mm |
| maximal/minimalny track width | 3350/2350 mm |

**Fig. 1.** Vehicle mounted on a load bearing chassis dynamometer.

The tested vehicle was adjusted through modification of the engine electronic system involving changing the manufacturer's computer software of the drive unit. Introduction of changes into the fuel injection controller was supposed to discover whether and how the vehicle performance parameters undergo change. Due to the specificity of the vehicle electronic system it was necessary to remove the deck computer each time and introduce changes into the software on a special test stand which is presented in figure 2. Before introduction of modifications to the computer software, a diagnostic measurement was performed to determine the possibilities and advisability of improving the engine performance parameters.

**Fig. 2.** Test stand used for modifications.

Changes in the fuel injection controller settings were introduced according to a pre-set experiment plan. Modifications of the vehicle computer involved increasing the fuel doses and the amount of air supply. The tests covered five settings of the fuel injection controller which are presented in table 4. Figure 3 shows an exemplary characteristics of fuel injection for an increased fuel dose.

**Table 4.** Fuel injection controller settings.

| Setting number | description |
|---|---|
| I | Manufacturer's settings |
| II | Fuel dose increased by 2% and air charge increased by 50 hpa |
| III | Fuel dose increased by 4% and air charge increased by 50 hpa |
| IV | Fuel dose increased by 6% and air charge increased by 50 hpa |
| V | Fuel dose increased by 6% and air charge increased by 150 hpa |

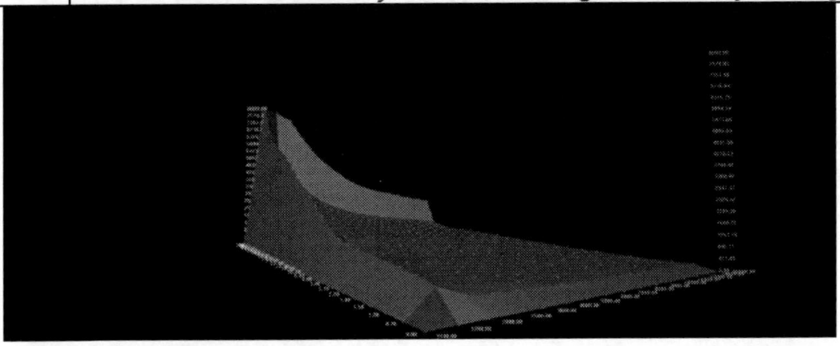

**Fig. 2.** Characteristics of fuel injection for an increased fuel dose.

Power and torque tests for different settings of fuel injection controller were repeated 30 times. Results of the tests were subjected to statistical analysis with the use of Statistica program.

## 3 Results

The test results have proven that for the manufacturer's settings, the maximal power reached by the drive unit was 102 KM, at 3600 rev/min and maximal torque 232 Nm, at 2400 rev/min. Increasing the dose of fuel and air to be supplied to the improvement in the engine performance parameters for all rotational speeds. The differences of power test results were statistically significant and involved changes in the range from 0 KM to 20 KM. The biggest differences were reported for rotational speeds in the range from 2900 rev/min do 3400 rev/min. For low rotational speeds (0-2500 rev/min) the differences were statistically insignificant. Increasing the fuel dose by 2% did not cause significant growth in the engine power input as compared to the drive unit standard performance parameters. No sooner than after increasing the fuel dose by 4%, and the air dose by 50 hPa, the power increase was statistically significant (for $p=0.05$). Further increasing the fuel dose did not cause a significant power increase. Increasing the air dose by 150 hPa, with a higher fuel dose 6%, had also an influence on the power improvement. Setting III was found to be the best setting of the fuel injection controller (increasing the fuel dose by 4% and air load by 50 hPa). The diagram of power in dependence on the engine rotational speed is shown in figure 3.

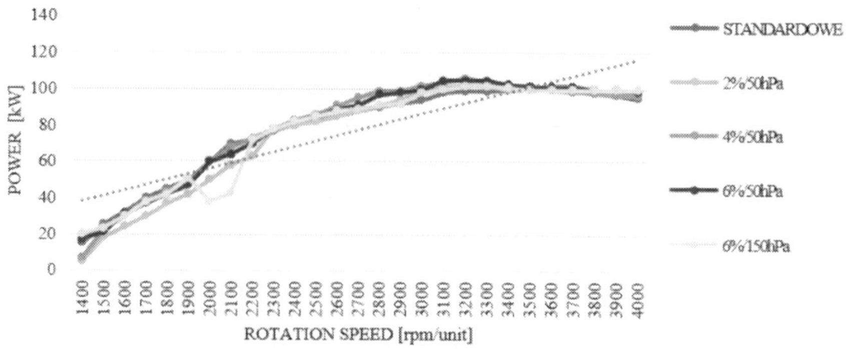


**Fig. 3.** Diagram of dependence between the engine power and rotational speed for the analyzed fuel injection controller setting changes.

The tests of torque yielded significant statistical differences which varied within the range from 10 Nm to 40 Nm. The most significant differences in rotational speed was found for from 2100 rev/min to 2700 rev/min. In the case of low rotational speeds (0-2000 rev/min) differences were not statistically significant. Increasing the fuel dose by 2% did not result in significant torque increase as compared to the engine standard parameters. It was the fuel dose increase by 6%, for increased air charge by 50 hPa that resulted in statistically significant (for p=0,05) torque increase. Application of air chargé increased by 150 hP for an increased fuel dose by 6%, did not have a significant impact on the engine torque improvement. Setting IV (increasing the fuel dose by 6% and air charge by 50 hPa) ) was found to be the best fuel injection controller setting for all the analyzed rotational speeds. A diagram of torque in dependence on the engine rotational speed is presented in figure 4.

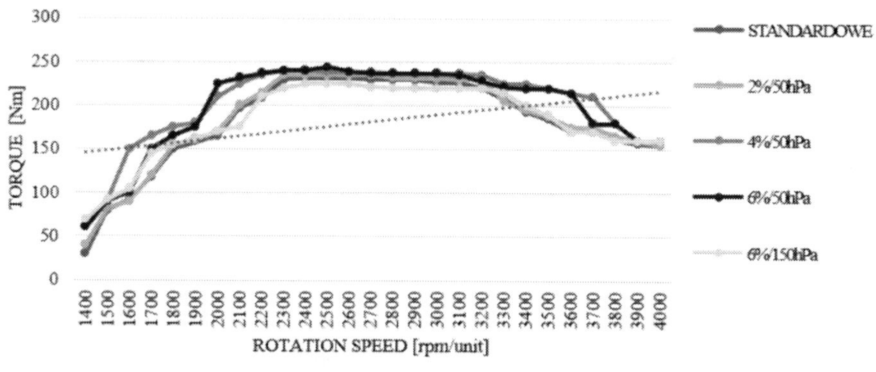


**Fig. 4.** A diagram of dependence of torque on rotational speed for the analyzed fuel injection controller setting changes.

The test of the engine power output and torque depending on the engine rotational speed and for variable fuel injection settings were conducted of an engine powered with pure diesel fuel (with no biocomponent additives).

## 4 Conclusions

Modifications of the fuel injection controller standard settings contribute to the engine power and torque increase which has been proven in many research centers.

The test results have revealed that increasing the fuel supply to the engine and increasing the air charge allowed to provide the drive unit of a motor vehicle with higher power and torque.

The power increase was found to be up to 20 KM, which was recorded for all rotational speeds of the engine. The biggest differences in measurement results were observed for rotational speeds in the range of 2900-3400 rev/min. Setting III was the optimal one for the fuel injection controller of the analyzed research object for power measurement (fuel dose increased by 4% and air load by 50 hPa). Further increasing the fuel dose and air charge caused no increase in the drive unit power.

Differences varying within the range from 10 Nm to 40 Nm were observed when measuring the torque for different fuel injection controller settings. The most significant differences were found for rotational speed of 2100- 2700 rev/min. Setting IV (increasing the fuel dose by 6% and air load by 50 hPa) was found to be the optimal setting of the fuel injection controller .

The analysis of power and torque test results indicates that modifications of the drive unit computer software provided the engine with better performance parameters.

## References

1. Lio H., Yu J., Xu J., Bao X., Identification of key oil refining technologies for China National Petroleum Co. (CNPC), Energy Policy, **35** (2007)
2. Castelo Branco D., Gomes G., Szklo A., Challenges and technological opportunities for the oil refining industry: A Brazilian refinery case, **38** (2010)
3. Demirbas A., Biofuels, securing the planet future Energy needs, London, Sringer Verlag Limited, **50** (2009)
4. Demirbas A., Importance of biodiesel as transportation fuel, Energy Policy, **35** (2007)
5. Pająk M., Muślewski Ł., Landowski B., Grządziela A., Fuzzy identification of the reliability state of the mine detecting ship propulsion system, Polish Maritime Research, **1** (2019)
6. PN-EN 590+A1:2017, Fuels for motor vehicles Diesel oils -Requirements and research methods (2018)
7. Mysłowski J., Development tendencies of self-ignition engines, Buses, **6** (2006)
8. Kuszewski G., Świątek D., Gutek D., Modification of a passenger computer software and its impact on the performance and operation economy of a drive unit, Buses, **12** (2017)
9. Prajkowski K., Impact of an external, so called box, module on gases composition of the ROVER 2.0 CDTI engine, Journal of Kones, **20** (2013)
10. Boguta A., Electronic modification of a self-ignition car engine performance parameters, MOTROL, **13** (2011)
11. Arcaklioglu E., Celikten I., A diesel engine performance and exhaust emissions, Applied Energy, **80** (2005)
12. Kaszkowiak J., Markiewicz M., Krzaczek P., Impact of the application of bioesters' addition to diesel oil on the course of turning moment and pwer within the scope of low rotational speed at variable settings of fuel injection, International Conference on Trends in Agricultural Engineering (2019)
13. Kołodziej E., Skrzyniowski A., Badania diagnostyczne samochodów na hamowni podwoziowej Autodyn 30, Mechanika, **10** (2012)
14. Kralik M., Jablonicky J., Nikolov M., Monitoring of NOx emission at selected diesel engine, University of Ruse „Angel Kanchev" (2015)
15. Michalski R., Gonera J., Janulin M., Wyznaczanie charakterystyki zewnętrznej ciągnika kołowego z wykorzystaniem przenośnej hamowni intercyjnej, Postępy Nauki i Techniki, **14** (2012)

# Diagnostics of Solar Cells

*Valeriy* Martynyuk[1,*], *Juliy* Boiko[1], *Marcin* Łukasiewicz[2], *Ewa* Kuliś[2] and *Janusz* Musiał[2]

[1]Programming and Computer, Telecommunication Systems Faculty in Khmelnytsky National University, Department of Telecommunication and Computer Integrated Technology, 29016, 11, Instytuts'ka str., Khmelnitsky, Ukraine
[2]Faculty of Mechanical Engineering UTP in Bydgoszcz, Department of Vehicle Engineering, Al. Prof. S.Kaliskiego 7, 85-796 Bydgoszcz, Poland

**Abstract.** The paper represents the mathematical model for diagnostics of solar cell. The research objectives are the problem of determining a solar cell technical condition during its operation. The solar cell diagnostics is based on the mathematical model of solar cells. The single-diode solar cell model is characterized by a slight deviation of the theoretically calculated characteristics from the characteristics of the real solar cell, one of the reasons being the complexity of the accurate measurement of the series resistance. The single-diode solar cell model uses the current and voltage ratio in the form of an implicit function and it cannot be solved directly. For its solution it is necessary to use numerical methods. This is main disadvantage of the single-diode solar cell model. The methodological approach to increasing the reliability of the solar cell diagnostic has been proposed, in terms of multi-parameter the solar cell diagnostic by applying the solar cell impedance model.

## 1 Introduction

The problem of finding alternative energy sources has become particularly urgent over the last few decades. The reasons for this are the lack of extraction of fossil fuels, primarily oil and gas, as well as environmental pollution caused by their use. On the other hand, the energy of solar radiation is a free, almost everywhere available resource, which allows generating environmentally friendly electricity without the least damage to the environment. In this context, solar energy is one of the most promising sectors of rapidly developing alternative energy.

Today, semiconductor devices with p-n-junctions, which are called solar or photovoltaic (PV) cells, are widely used in the field of solar energy. In such devices, due to the internal photoelectric effect under the influence of solar radiation, there is a redistribution of charges that results in the appearance of electromotive force (EMF), which makes it possible to directly convert the energy of sun light into electrical energy [1, 2].

The quality and reliability of solar cells are largely determined by diagnostics support at all stages of their life cycle. At the production stages, where it is crucial to ensure zero-defect manufacturing technology and labour for producing solar cells, the volume of control and diagnostic operations reaches 50% of the total labour intensity of their production. A characteristic problem of diagnosing solar cell manufacturing defects under the conditions of their mass production is that the later the defects are detected, the more costs are necessary to localize the place of their emergence [3, 4].

Hence, diagnostic support will be effective only when the high reliability values are ensured at minimum cost. At the operation stage, the purpose of diagnosing a solar cell is to evaluate its technical condition, to determine the reasons of its operation failures, to predict changes in a solar cell technical condition and determine the reasons for such changes or

* Corresponding author: martynyuk.valeriy@gmail.com

time intervals after which processes causing undesirable changes in a solar cell technical condition might begin.

At present, the renewable energy sources are the most important issues of modern energetics that is described in [5]. Thus, the issues of developing methods and means of solar cell diagnostics have become urgent.

Solar cell diagnostics is an information procedure the purpose of which is to reflect its technical condition in the form of a conclusion about the nature and significance of this condition. Such a conclusion regarding the diagnostic results is called a diagnosis [3].

Technical condition of a solar cell is a condition that is characterized at a certain time, under certain environmental conditions by the values of solar cell parameters set by its data sheet. According to the existing practice, it has been confirmed to define such mutually exclusive incompatible pairs of technical conditions for technical products: serviceable – unserviceable, operable – faulty.

For the serviceable – unserviceable conditions the basis of their comparison is compliance of all solar cell parameters with the requirements of normative data sheet. And if at least one of these parameters, regardless of its impact on the solar cell ability to perform its functions, will not fulfill these requirements, the technical condition is considered unserviceable. And when all the parameters meet all the requirements of proper documentation, the technical condition is serviceable.

A solar cell is in an operable technical condition if its relevant diagnostic parameters are within the limits where working functions can be performed. When the diagnostic parameters go beyond the specified limits, then the solar cell can no longer fully perform its functions, and it has to be considered faulty.

## 2 Solar cell mathematical model

The single-diode solar cell model received widespread application [1, 2]. It provides an opportunity to perform calculations of the illuminated current-voltage characteristic of a separate panel as well as an array (string) of solar cells connected in series at different values of solar irradiance and the surface temperature of the panel.

An equivalent circuit that corresponds to a single-diode solar cell model is shown in fig. 1. The circuit consists of a current source, one diode, a series and a parallel resistance. The typical current-voltage and power-voltage characteristics of solar cell are shown in fig. 2. The single-diode model in fig. 1 is characterized by a slight deviation of the theoretically calculated characteristics from the characteristics of the real solar cell, one of the reasons being the complexity of the accurate measurement of the series resistance in a solar cell model [2].

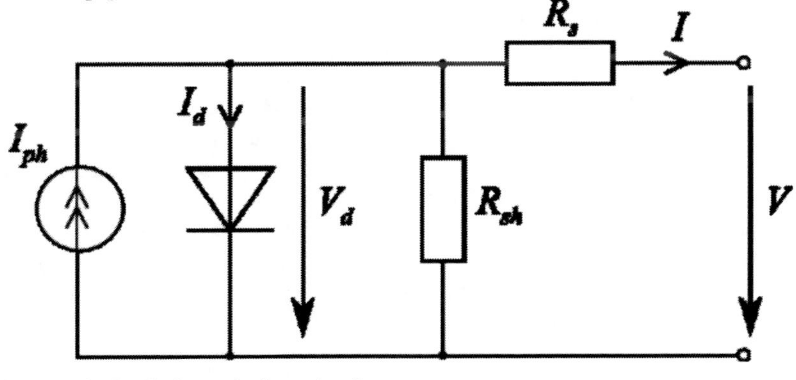


**Fig. 1.** Solar cell single-diode equivalent circuit.

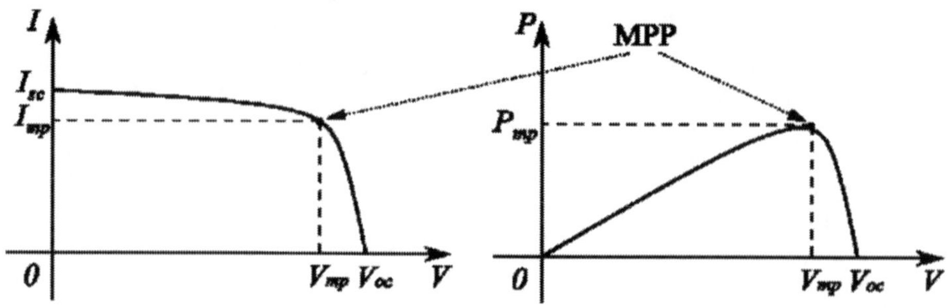


**Fig. 2.** The typical I-V and P-V curves of PV cell.

There are more complicated models that can provide more accurate description of the solar cell characteristics [4]. For example, it is possible to use the model with two diodes [5], one of which simulates the effect of recombination of charges. According to [5], the two-diode model quite accurately describes crystalline silicon solar cells, but for thin-film solar cells the accuracy of approximation may have some limitations. On the other hand, the use of the two-diode model leads to an increase in the number of unknown parameters, which complicates the calculation of the model. So, in this paper the single-diode model of the solar cell is preferred.

The question of choosing an adequate mathematical model for calculating the I-V curves of the solar cell is considered in the following papers [2]. In the paper [5] the problems of simulating I-V curves of solar cells are considered. The authors describe in detail different models of solar cell, analyze the conditions of their application and the accuracy of I-V curves simulation, depending on the solar irradiance and temperature. The single-diode and two-diode solar cell models are compared in works [4-7].

The input data for calculating the solar cell parameters are the value of its I-V curve at three characteristic points corresponding to the short circuit, open circuit and maximum power point (MPP) at the solar irradiance $G = 1$ kW/m$^2$ and solar cell surface temperature $T$=25 °C. These values of irradiance and temperature correspond to the so-called standard test conditions (STC).

The values of short-circuit current $I_{sc0}$, open-circuit voltage $U_{sc0}$, as well as current and voltage at the maximum power point $P_{mp0}$ under STC, are usually provided by solar panel manufacturers together with the corresponding temperature coefficients. Series resistance $R_s$, parallel resistance $R_{sh}$, photocurrent $I_{ph}$, as well as saturation current $I_s$ and ideality factor for a diode $m$ remain unknown [8, 9].

There are quite a lot of methods for calculating the unknown solar cell parameters, which differ in the approaches to the solution of the I-V equation given in an implicit form. For example, paper [4] describes a method based on the decomposition of the I-V equation for a solar cell into the Taylor series. In paper [6], the authors present a methodology for calculating solar module parameters based on the datasheet information. Paper [6] is devoted to the development of a solar cell model in the Matlab/Simulink software environment. Paper [7] presents a method for calculating series and parallel resistances of single-diode solar cell model on the basis of I-V curve analysis.

In this paper, a detailed description of the procedure for calculating the parameters of a single-diode solar panel model is provided. The algorithm for calculating series and parallel resistances of the solar cell equivalent circuit is presented. On the basis of the obtained ratios, a simulation model of solar cell in Matlab/Simulink software environment was developed.

Basing on Kirchhoff's laws, we can write the following expression for the circuit in fig. 1:

$$I = I_{ph} - I_d - \frac{V + IR_s}{R_{sh}} \quad (1)$$

where: $I$ and $V$ are solar cell output current and voltage; $I_{ph}$ is photocurrent; $I_d$ is current flowing through the diode; $R_s$ is series resistance; $R_{sh}$ is parallel (shunt) resistance.

The expression for the current $I_d$ can be obtained on the basis of the ideal p-n-junction equation [2-5]. For the array consisting of $N_s$ PV cells connected in series, current $I_d$ is defined by the following ratio:

$$I_d = I_s \left[ \exp\left( \frac{V_d}{mV_T} \right) - 1 \right] \quad (2)$$

where: $I_s$ is reverse saturation current; $I_d = V + IR_s$ is voltage at the diode; $V_T$ is thermal voltage; $m$ is diode ideality factor.

The value of the thermal voltage $V_T$ is defined by the following expression:

$$V_T = \frac{N_s kT}{q} \quad (3)$$

where: $N_s$ is the number of solar cells connected in series; $T$ is absolute p-n-junction temperature; $k$ is Boltzmann constant; $q$ – elementary charge equal to the module of electron charge.

Thus, taking to account (1) - (3), the analytical expression for the I-V curve of an array of $N_s$ solar cells connected in series is the following:

$$I = I_{ph} - I_s \left[ \exp\left( \frac{V + IR_s}{mV_T} \right) - 1 \right] - \frac{V + IR_s}{R_{sh}} \quad (4)$$

The model, represented by expression (4), allows us to develop theoretical I-V curves of solar panels with an accuracy that is sufficient for engineering calculations and computer simulation [2].

It includes 5 unknown parameters that have to be found in different ways: $I_{ph}$, $I_s$, $m$, $R_s$ and $R_{sh}$. There are many methods for defining these parameters, most of which are based on an assumption that $R_s$ takes low values, but $R_{sh}$, on the contrary, is quite large.

Thus, assigning $R_s = 0$ and $R_{sh} = \infty$ in expression (4), we can get a simplified expression for I-V characteristic:

$$I = I_{ph} - I_s \left[ \exp\left( \frac{V}{mV_T} \right) - 1 \right] \quad (5)$$

Expression (5) corresponds to the model of the ideal solar cell. It allows representing I-V characteristic in the form of explicit function $I = I(V)$, but with less precision than in expression (4) [4, 5].

The coefficient $m$ is different for each material a solar cell is made of. For example, for the solar cells made of polycrystalline silicon $m = 1.3$ [9].

Let's consider I-V curve of solar cell at the point (0, $I_{sc0}$), which corresponds to short circuit (fig. 2).

Substituting these values into equation (5), we obtain the approximate value of the photocurrent under the STC:

$$I_{ph0} \approx I_{sc0} \quad (6)$$

where: $I_{sc0}$ is short-circuit current under the test conditions.

Let's represent the exact photocurrent value $I_{ph0}$ by the following expression:

$$I_{ph0} = \alpha I_{sc0} \tag{7}$$

where: $\alpha$ is the coefficient of proportionality.

The actual photocurrent value $I_{ph}$ depends on the temperature of the panel and solar irradiance, and, taking into account (6), is defined by the following ratio [8, 9, 10]:

$$I_{ph} = \frac{G}{G_o}\left[1 - K_I\left(T - T_o\right)\right]I_{ph0} \tag{8}$$

where: $G_0$=1 kW/m² is the solar irradiance under the STC; $T_0$=25°C is the solar cell temperature under the STC; $K_I$ is the temperature coefficient of short circuit.

Let's consider I-V curve at the point ($V_{oc0}$, 0), which corresponds to solar cell open circuit (fig. 2). Substituting these values into equation (5) and taking into account (6) we obtain the expression for the value of the diode saturation current under the STC:

$$I_{s0} = \frac{I_{sc0}}{\exp\left(\frac{V_{oc0}}{mV_{To}}\right) - 1} \tag{9}$$

where: $V_{oc0}$ and $V_{T0}$ are open-circuit voltage and thermal voltage under the STC correspondingly.

The dependence of the current $I_s$ on the temperature can be expressed with the help of temperature coefficients of short circuit and open circuit [8] according to the following ratio:

$$I_s = \frac{I_{sc0}\left(1 + K_I \Delta T\right)}{\exp\left(\frac{V_{oc0}\left(1 + K_V \Delta T\right)}{mV_T}\right) - 1} \tag{10}$$

where: $\Delta T = T - T_0$; $K_V$ is open-circuit temperature coefficient.

For the mathematical model (4), resistances $R_s$ and $R_{sh}$ remain unknown. They can be defined on the basis of the equality of the experimental and theoretical maximum power values.

It is known that the solar cell output power is a product of output voltage $V$ and current $I$:

$$P = V \cdot I \tag{11}$$

where $I$ and $V$ are PV panel output current and voltage correspondingly.

Let's substitute $I = I_{mp}$ and $V = V_{mp}$ into the equation (4) and, taking into account expression (11), write the expression which represents the ratio between $R_{sh}$ and $R_s$:

$$R_{sh} = \frac{V_{mp} + I_{mp}R_s}{I_{ph} - I_s\left[\exp\left(\frac{V_{mp} + I_{mp}R_s}{mV_T}\right) - 1\right] - I_{mp}} \tag{12}$$

From equation (12), for some $R_s$ we can find the corresponding value of $R_{sh}$, at which the theoretical I-V curve will pass through the experimental MPP. Gradually increasing the value of $R_s$, starting from $R_s = 0$, it's possible to find such a pair of [ $R_s$, $R_{sh}$ ] at which the theoretical MPP will be equal (with the known error) to the experimental one. To achieve this, for each pair of [ $R_s$, $R_{sh}$ ] from equations (4) and (12) we should find the maximum power value $P$ which corresponds to the theoretical I-V curve. Thus, on each iteration of the algorithm the values of $R_s$ and $R_{sh}$ are approaching the optimal ones, that provide the best possible accuracy of the model.

The initial value of resistance $R_{sh}$ can be defined according to the following expression [8]:

$$R_{sh0} = \frac{V_{mp}}{I_{sc0} - I_{mp}} - \frac{V_{oc0} - V_{mp}}{I_{mp}} \tag{13}$$

where: $I_{mp}$ and $V_{mp}$ are current and voltage corresponding to the experimental MPP.

As shown in expression (6), $I_{ph0} \neq I_{sc0}$. The photocurrent value $I_{ph0}$ under the STC can be expressed with higher accuracy if to put $\alpha = 1 + \frac{R_s}{R_{sh}}$ into expression (7). So, the final expression for the photocurrent $I_{ph0}$ looks like this:

$$I_{ph0} = \left(1 + \frac{R_s}{R_{sh}}\right) I_{sc0} \tag{14}$$

Since equation (4) sets the current and voltage ratio in the form of an implicit function $f = f(I,V)$, it cannot be solved directly. For its solution it is necessary to use numerical methods [7].

## 3 Solar cell diagnostics system

In terms of electrical engineering, a solar cell is an active bi-polarity with frequency dispersion of its electrical parameters. According to the theory of mathematical modeling, any mathematical model can be represented graphically and analytically.

Proceeding from this, under the impedance mathematical model of the solar cell in graphical form, we mean its equivalent circuit on alternating current. In analytical form, the impedance mathematical model of the solar panel is the analytical expression of its impedance, which combines the parameters of the equivalent circuit elements with the electrical parameters of the solar panel.

To determine the solar cell parameters we conducted the solar cell impedance measurements by means of the solar cell diagnostic system which is shown in fig. 3.

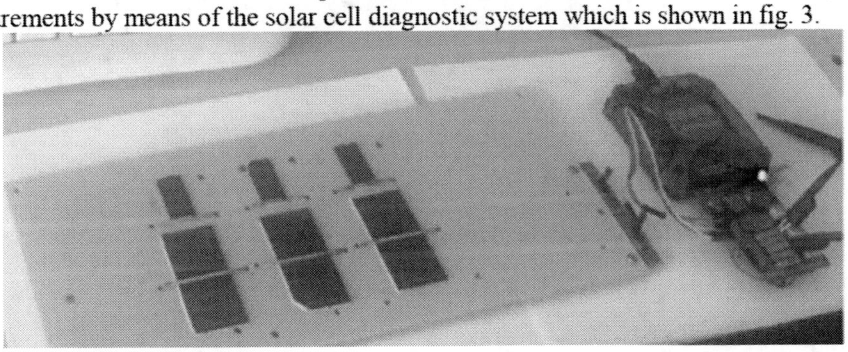

**Fig. 3.** Solar cell diagnostic system.

Structurally, the solar cell diagnostic system consists of an Analog Discovery 2 device produced by the US Digilent which is shown in fig. 4, a. To measure the impedance of solar cells, the specialized Wave Forms software by the US Digilent was used as shown in fig. 4, b.

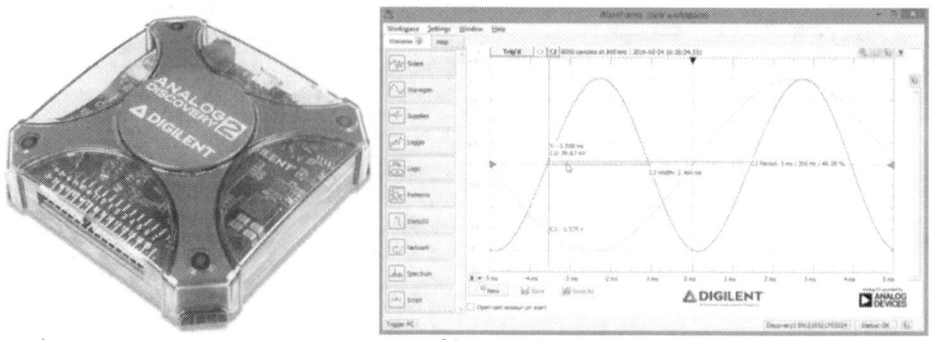


a) b)

**Fig. 4.** The picture of Analog Discovery 2 and Wave Forms software

Analog Discovery 2 is connected to an adapter assembled on a model board and laboratory bench as shown in fig. 5.

Digital outputs of the Analog Discovery 2 are used to connect one of the reference resistances needed to measure current through a solar cell. The solar cell data acquisition system allows estimating impedance in the solar cell from 1 Ω to 1 MΩ in the frequency range from 1 Hz to 1 MHz with a regular frequency distribution within the range.

The experimental measurements of impedance in the solar cell performed by the authors show that its experimental Nyquist plot has the form of two semicircles, as shown in fig. 6. The large semicircle corresponds to the barrier capacitance $C_b$ of the solar cell, the effect of which is observed at high frequencies (from 100 Hz to 1 MHz).

**Fig. 5.** The picture of adapter.

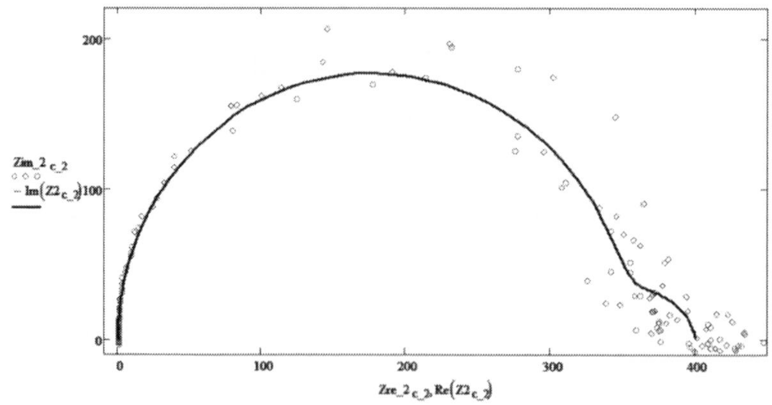


**Fig. 6.** Experimental and theoretical Nyquist plots of the solar cell.

The small semicircle corresponds to diffusion capacitance $C_d$ and frequency-dependent conductivity the influence of which is observed at low frequencies (from 1 Hz to 100 Hz). By analysing the experimental Nyquist plot as shown in fig. 6 we can conclude that the small semicircle is substantially deformed by the loss of electric energy and is under the influence of noise obstruction. Such a character of the Nyquist plot corresponds to the fractional capacitor with significant losses of electric energy [7, 8].

Thus, the authors suggest an impedance mathematical model of the solar cell, related to the expression of impedance (15).

$$Z(s) = R_s + \frac{R_b}{1+sC_bR_b} + \frac{R_d}{1+s^{\alpha}C_dR_d} \tag{15}$$

The adequacy of the suggested impedance mathematical model of the solar cell is confirmed by the convergence of the experimental and theoretical Nyquist plots, which are shown in fig. 6.

By analysing the expression (15) of impedance the authors propose the solar cell equivalent circuit as shown in fig. 7. The solar cell equivalent circuit consisting of series resistance $R_s$, barrier capacitance $C_b$, barrier resistance $R_b$, diffusion resistance $R_d$ and diffusion capacitance $C_d$ of the fractional order.

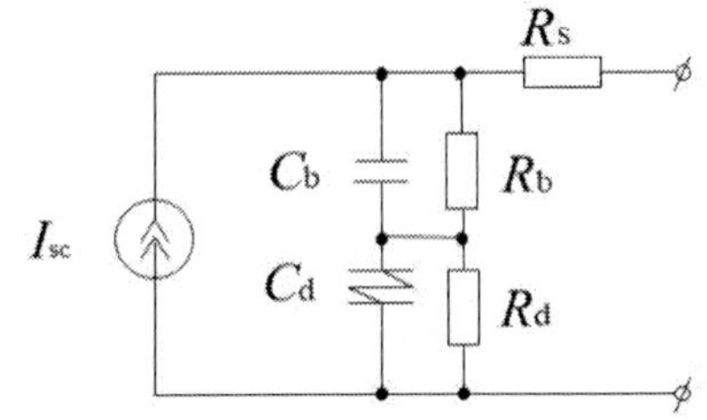


**Fig. 8.** The impedance solar cell equivalent circuit.

As a result of the approximation of the experimental Nyquist plot the following values of the parameters of the improved solar cell equivalent circuit have been obtained: $R_s$=0,1 Ω, $C_b$=200 nF, $R_b$=350 Ω, $C_d$=100 μF, $R_d$=50 Ω, α=0,9.

Next, two hypotheses are put forward.

1. The basic hypothesis $H_0$ (the solar cell is working):

$$\begin{aligned} &Rs \in [Rs_{\min}, Rs_{\max}]\ ; \\ &Rb \in [Rb_{\min}, Rb_{\max}]\ ; \\ &Rd \in [Rd_{\min}, Rd_{\max}]\ ; \\ &Cd \in [Cd_{\min}, Cd_{\max}]\ ; \\ &\alpha \in [\alpha_{\min}, \alpha_{\max}]\ . \end{aligned} \tag{16}$$

where: $Rs_{min}$, $Rs_{max}$; $Rb_{min}$, $Rb_{max}$; $Rd_{min}$, $Rd_{max}$; $Cd_{min}$, $Cd_{max}$ and $\alpha_{min}$, $\alpha_{max}$ are limits of permissible interval for Rs, Rb; Rd, Cd and α respectively.

2. Alternative hypothesis $H_1$ (the solar cell is defective):

$$\begin{aligned} &Rs \notin [Rs_{\min}, Rs_{\max}]\ ; \\ &Rb \notin [Rb_{\min}, Rb_{\max}]\ ; \\ &Rd \notin [Rd_{\min}, Rd_{\max}]\ ; \\ &Cd \notin [Cd_{\min}, Cd_{\max}]\ ; \\ &\alpha \notin [\alpha_{\min}, \alpha_{\max}]\ . \end{aligned} \tag{17}$$

The diagnostic reliability of the solar cell is defined as the product of the methodological and the instrumental diagnostic reliabilities of the solar cell. The methodological diagnostic reliability of the solar cell is defined by the ratio between the number of exponents characterizing the technical condition of the object of diagnostic and included to the mathematical model of diagnostics and the total number of exponents characterizing the technical condition of the solar cell.

The instrumental diagnostic reliability of the solar cells is defined by the probabilities of errors of the first and second kind, $\alpha$ and $\beta$ respectively. Increase in the diagnostic reliability of the solar cell is possible when using the solar cell characteristics that contain information on the nominal values of technical condition of the solar cell in the range of changes in nominal voltages in the operation current range. Such characteristics include nonlinear resistance $Z(s,u)$ and nonlinear transient resistance $z(t,u)$ of the solar cells, related to each other by inverse and direct Laplace transforms [9-11]:

$$Z(s,u) = L^{-1}\left[z(t,u)\right]; \tag{18}$$

$$z(t,u) = L\left[Z(s,u)\right], \tag{19}$$

where: $s = j\omega$ is complex frequency.

## 7 Conclusions

Solar cells are the basis of solar power plants. Each year, the production of solar cells is increasing, so their diagnostics is very important to the electric energy production. The solar cell diagnostics is based on the mathematical model of solar cells. The single-diode solar cell model is characterized by a slight deviation of the theoretically calculated characteristics from the characteristics of the real solar cell, one of the reasons being the complexity of the accurate measurement of the series resistance. The single-diode solar cell model uses the current and voltage ratio in the form of an implicit function and it cannot be solved directly. For its solution it is necessary to use numerical methods. This is main disadvantage of the ingle-diode solar cell model.

To solve this disadvantage the authors propose to use the impedance solar cell model. The impedance solar cell model is based on the series resistance $R_s$, barrier capacitance $C_b$, barrier resistance $R_b$, diffusion resistance $R_d$ and diffusion capacitance $C_d$ of the fractional order. These parameters were obtained from the solar cell experimental impedance data. The solar cell impedance was measured by means of the solar cell diagnostic system which is developed by authors.

## References

1. T. Ahmad, S. Sobhan, S. Performance Study of Photovoltaic Solar Cell. International Jour. of Research in Engineering and Technology, V. 4, pp. 12-18, (2015)
2. C. Carrero, J. Amador, S. Arnaltes. A Single Procedure for Helping PV Designers to Select Silicon PV Modules and Evaluate the Loss Resistances. Renewable Energy, V. 32, pp. 2579-2589, (2007)
3. V. Martynyuk, O. Eromenko, J. Boiko, T. Kałaczyński. Diagnostics of supercapacitors. MATEC Web Conf., **182**, 01009, (2018)
4. P. Arjyadharal, Ali S.M, J. Chitralekha. Analysis of Solar PV cell Performance with Changing Irradiance and Temperature. Inter. Jour. of Engineering and Computer Science, V. 2, I. 1, pp. 214-220, (2013)
5. M. G. Villalva, J. R. Gazoli, E. Ruppert F. Modeling and circuit-based simulation of photovoltaic arrays. Brazilian Jour. of Power Electronics, V. 14, no. 1, pp. 35-45, (2009)
6. H. Bellia, R. Youcef, M. Fatima. A detailed modeling of photovoltaic module using MATLAB. NRIAG Jour. of Astronomy and Geophysics, pp.53-61, (2014)

7. M. Ortigueira, J. Trujillo, V. Martynyuk, J. Coito. A generalized power series and its application in the inversion of transfer functions. Jour. of Signal Processing, V. 107, pp. 238-245, (2015)
8. V. De Santis, V. Martynyuk, A. Lampasi, M. Fedula, M. Ortigueira. Fractional-order circuit models of the human body impedance for compliance tests against contact currents. AEU – Inter. Jour. of Electr. and Communcations, V. 78, pp. 238-244, (2017)
9. V. Martynyuk, D. Makaryshkin, J. Boyko, "Frequency domain analysis for electrochemical supercapacitors", in 2007 Proc. 15th IMEKO Symposium on Novelties in Electrical Measurements and Instrument. in Parallel with the 12th Workshop on ADC Modelling and Testing, Iasi, pp. 5, (2007)
10. V. Martynyuk, D. Makaryshkin, J. Boyko, "Electrochemical supercapacitor time domain analysis by means of multi-channel measurement system", in 2007 Proc. 15th IMEKO Symposium on Novelties in Electrical Measurements and Instrument. in Parallel with the 12th Workshop on ADC Modelling and Testing, Iasi, 2007, Vol. I, pp. 207–211, (2007)
11. T. Kałaczyński, M. Łukasiewicz, J. Musiał, R. Polasik, M. Szczutkowski, N. Dluhunovych, J. Wilczarska, T. Kasprowicz, *Analysis of the diagnostic potential research thermovision in the technical state of combustion engine injectors assessment*, Engineering Mechanics, 1805-8248, 357-360, (2018)

# Diagnostics of Corrosion Protection Potential for Electric Water Heaters

*Valeriy* Martynyuk[1*], *Michał* Liss[2] and *Joanna* Wilczarska[2]

[1]Programming and Computer, Telecommunication Systems Faculty in Khmelnytsky National University, Department of Telecommunication and Computer Integrated Technology, 29016, 11, Instytuts'ka str., Khmelnitsky, Ukraine
[2]Faculty of Mechanical Engineering UTP in Bydgoszcz, Department of Vehicle Engineering, Al. Prof. S. Kaliskiego 7, 85-796 Bydgoszcz, Poland

**Abstract.** The paper represents the mathematical model for diagnostics of the corrosion protection potential for the electric water heaters (EWH). The research objectives are the problem of determining the EWH corrosion protection technical condition during its operation. The EWH corrosion protection diagnostics is based on the mathematical model of the corrosion protection potential for EWH. The model is based on the nonlinear equivalent circuit which is modelling the nonlinear impedance of the corrosion cathodic protection. In the development of the mathematical model the authors carried out experimental measurements of the impedance for the EWH body fragment. The obtained experimental results show that mathematical model includes the fractional integrating elements. By using of the experimental results the expressions of the impedance, the impulse response is obtained. The methodological approach to increasing the reliability of the EWH corrosion protection diagnostics has been proposed, in terms of the multi-parameter diagnostics, by applying the EWH corrosion protection impedance model.

## 1 Introduction

Protection of metals from corrosion is one of the most important national economic problems, since losses from corrosion in industry, transport and other sectors of the economy are comparable to the costs of developing the largest industries. Technological and household systems for storing and processing industrial and drinking water are in a number of metal structures with increased labour input in manufacturing. A significant part of the products is manufactured using low alloy carbon steel with an enameled surface.

The use of alloyed stainless steel increases the cost of production, reduces its market quality. In addition, the increased content of chlorine ions in water and aqueous electrolytes reduces the corrosion resistance and can cause local or point destruction of the surface of alloys. In view of this, instruments and assemblies made of stainless steel should also be equipped with corrosion protection [1].

The rate of metal corrosion in water, soil and aqueous solutions depends on the potential $E$ of the material in the environment. When the potential is shifted towards the negative values, the corrosion rate decreases, which is achieved by means of direct electric current applied through the anode system to the metal surface of the protected object.

The material wear rate defined by the standards should be less than 0.01 mm per year. This wear rate provides the intended service life of the object, at which the influence of any corrosive wear is not manifested.

Electric water heaters are also called boilers. Their internal volume is filled with water, which is heated in the process of their work. The inner working surface of the electric water

* Corresponding author: martynyuk.valeriy@gmail.com

heater is in constant contact with the electrolyte, since ions of different metals are in the water. Thus, it can be concluded that the inner working surface of an electric water heater is exposed to electrochemical corrosion.

The cathodic protection of the metal surface, due to its manufacturability and the possibility of application in a wide range of environmental parameters, is by far the most common way of corrosion protection [2]. The cathodic protection is based on a shift of the metal potential towards negative values, while the rate of the anode reaction decreases in case of cathodic polarization, which causes a decrease in the corrosion rate.

Electric water heaters are one of the most common types of enameled household heating equipment [3]. According to EWH type, they are classified into: flow, storage (capacitive) and flow-storage. The design and operation of all capacitive EWH are almost the same as shown in fig. 1.

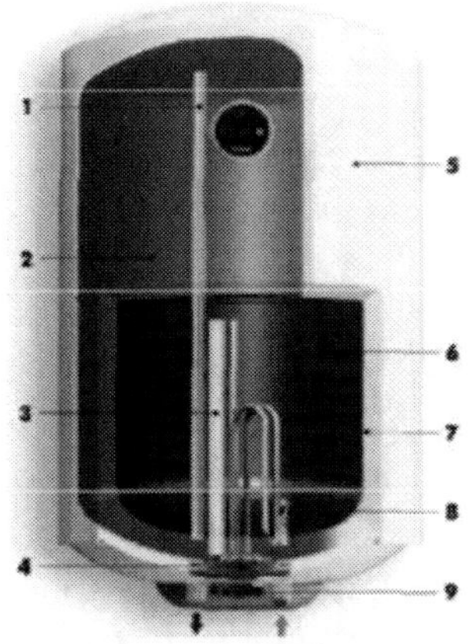


**Fig. 1.** EWH Design: 1– hot water outlet pipe; 2 – glass-enamel coated inner steel tank; 3 – magnesium anode; 4 – steel flange on which heating element, magnesium anode and thermostat are mounted; 5 – polymer-coated external steel tank; 6 – glass-enamel coating; 7 – polyurethane foam insulation; 8 – cold water supply pipe; 9 – thermostat.

EWH consists of inner and external steel tanks, the space between which is filled with insulating materials. Water heaters, a thermostat, and a safety valve are built into the inner tank.

The inner tank is made of low-carbon 1.5 – 2.0 mm sheet steel, and then coated with protective glass enamel. Upon contact of the inner steel tank with hot water in places of damage to the glass-enamel protective coating, intense corrosion of the metal surface occurs, which leads to destruction of the product.

The main reason for shortening the service life of a capacitive EWH leading to its failure is corrosion of its inner steel tank. Glass-enamel coating does not provide an EWH with complete protection from corrosion, because during its operation microcracks and other small defects can be formed due to cyclic thermal loads.

To avoid corrosion of defective coating areas of the EWH inner steel tank (not protected with glass-enamel coating), the cathodic protection method is used.

The main criteria of cathodic protection are the corrosion protection potential, as well as methods and devices for its diagnostics. Protective potential is the potential at which the metal dissolution rate takes an extremely low value, acceptable for these operating conditions.

## 2 Reliability of the EWH protection potential diagnostics

Reliability of diagnostics is the degree of objective conformity of the result of diagnostics to the actual technical condition of the corrosion protection potential, which is estimated by the probability of making the correct decision about the condition of the corrosion protection potential. The components of the general reliability of diagnostics are methodological and instrumental reliabilities of diagnostics [4]:

$$D_G = D_M \cdot D_I \quad (1)$$

where: $D_M$ is methodological reliability of diagnostics; $D_I$ is instrumental reliability of diagnostics.

Methodological reliability of diagnostics is calculated by the following formula:

$$D_M = \frac{N_1}{N} \quad (2)$$

where: $N_1$ is the number of exponents which characterize the technical condition of the supercapacitor and are included into the mathematical model of diagnostics;

N is total number of exponents which characterize the diagnostics of the EWH protection potential.

Instrumental reliability of diagnostics is defined by the probabilities of errors of the first and second kind respectively $\alpha$ i $\beta$ [5]:

$$D_I = 1 - \alpha - \beta \quad (3)$$

where: $\alpha$ is probability of errors of the first kind; $\beta$ is probability of errors of the second kind.

The overall reliability of diagnostics of the corrosion protection potential of EWH bodies $D_G$ may be increased by increasing its two components: methodological reliability of diagnostics $D_M$ and instrumental reliability of diagnostics $D_I$ .

Increase of the methodological reliability of diagnostics of the corrosion protection potential of EWH bodies $D_M$ is carried out in the process of diagnostics by determining a large number of parameters that affect the diagnostic object.

Increase of the instrumental reliability of diagnostics of the corrosion protection potential of EWH bodies $D_I$ is carried out by reducing the measurement error of controlled parameters.

## 3 Process of the EWH corrosion protection potential diagnostics

In general, the process of the EWH corrosion protection potential diagnostics comprises three main stages as shown in fig. 2.

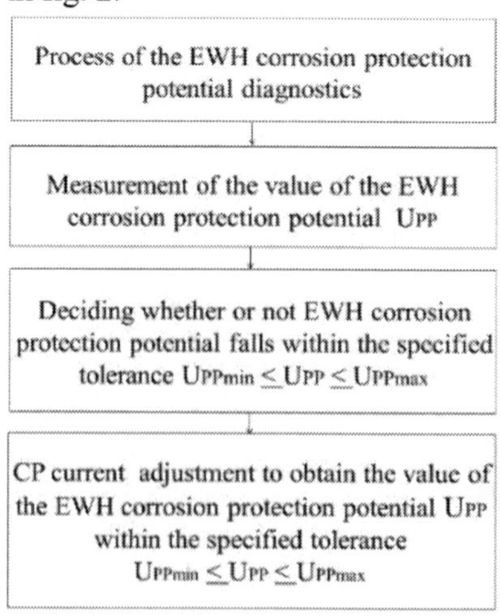


**Fig. 2.** The stages of the EWH corrosion protection potential diagnostics.

1. Measurement of the value of the EWH corrosion protection potential $U_{PP}$.

2. Deciding whether or not EWH corrosion protection potential falls within the specified tolerance $U_{PP\min} \le U_{PP} \le U_{PP\max}$. According to German standards DIN 50927-2013 the EWH corrosion protection is provided at the maximum value $U_{PP\max} = -540\,mV$. The minimum value of the protection potential $U_{PP\min} = -1{,}1V$ is limited by the voltage of water decomposition releasing hydrogen and oxygen and generating explosive gas. Generation of explosive gas is not permitted for safety reasons.

3. CP current $I_{CP}$ adjustment to obtain the value of the EWH corrosion protection potential $U_{PP}$ within the specified tolerance $U_{PP\min} \le U_{PP} \le U_{PP\max}$.

At the interface between the metal and the electrolyte solution, an electric double layer (EDL) appears [6-8]. Figure 3 presents an equivalent circuit of the protected from corrosion object (cathode) and the anode, located in a conductive medium (water).

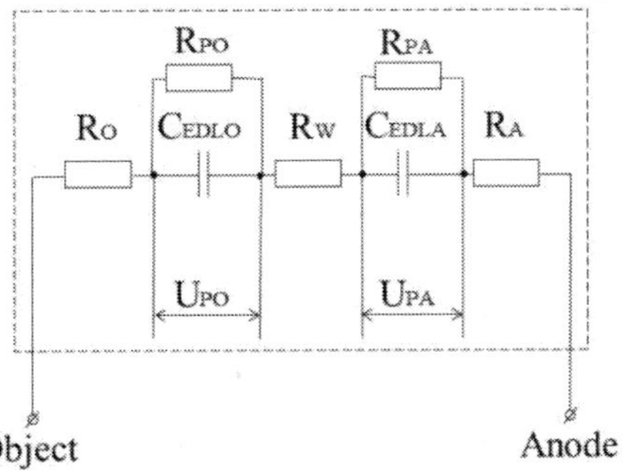


**Fig. 3.** The equivalent circuit of the object (cathode) and the anode, located in the conductive medium.

On the surface of the protected object (cathode), as well as on the surface of the anode, two EDL appear, which in fig. 3 are shown as two capacitors. Also, fig. 3 represents the active resistance $R_O$ and polarization resistance $R_{PO}$ of the protected object (cathode), the active resistance $R_A$ and polarization resistance $R_{PA}$ of the anode, which are connected through the active resistance of the conductive medium (water) $R_W$.

While developing an equivalent circuit (fig. 3), experimental measurements of the impedance of an EWH body fragment (cathode, fig. 4 a) in the electrolyte solution (tap water, fig. 4 b) and anode (fig. 4 c) were carried out.

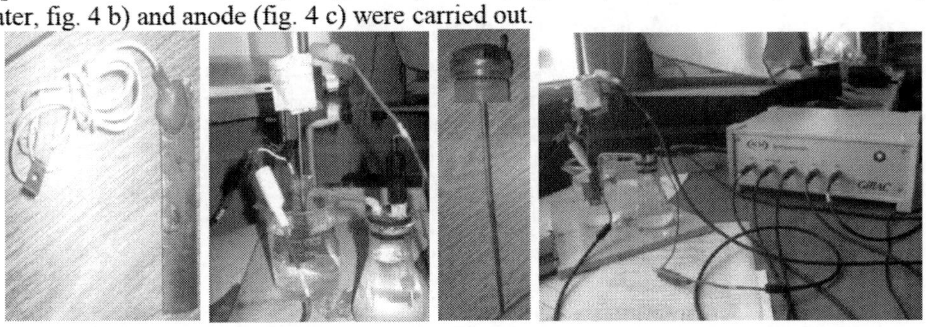

a) b) c) d)

**Fig. 4.** Pictures of the experimental setup components: a) an EWH body fragment (cathode); b) electrolyte solution (tap water); c) anode; d) experimental setup.

The measurements were carried out within the Department of Physicochemical Methods of Metal Corrosion Protection at Karpenko Physico-Mechanical Institute of the NAS of Ukraine (Lviv) using a GillAC impedance meter produced by ACM Instruments (Great Britain). A picture of the experimental setup is shown in fig. 4 d.

The adequacy of the equivalent circuit (fig. 3) is confirmed by the coincidence of the theoretical and experimental Nyquist diagrams, which are shown in fig. 5.

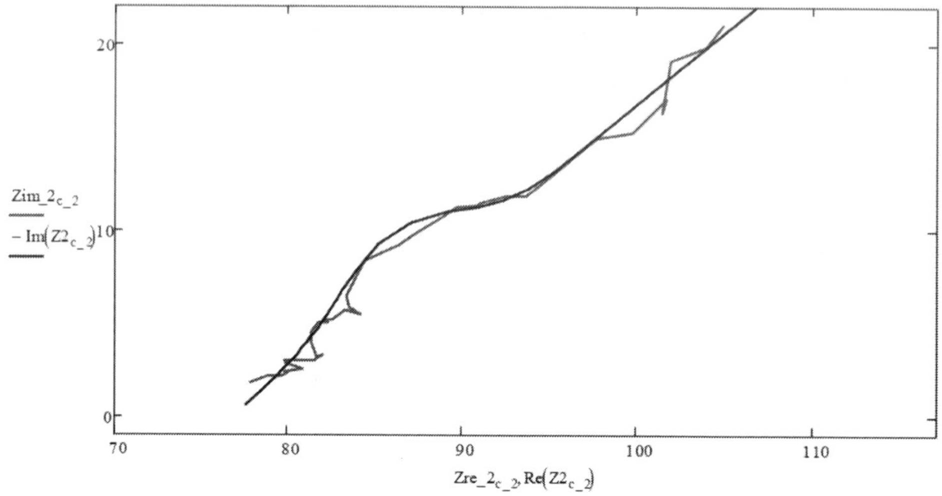


**Fig. 5.** Theoretical (blue curve) and experimental (red curve) Nyquist diagrams of the equivalent circuit of the EWH and anode, located in the conductive medium.

## 4 Mathematical model of EWH corrosion protection potential diagnostics

Approximation of the experimental Nyquist diagram by the theoretical Nyquist diagram is performed by the impedance function (4) in the frequency range from $F_{min}$=10 mHz to Fmax=187 Hz applying the least squares method (LSM).

$$Z(j\omega) = R_O + \frac{R_{PO}}{1+(j\omega)^{\alpha} C_{EDLO} R_{PO}} + R_W + \frac{R_{PA}}{1+(j\omega)^{\beta} C_{EDLA} R_{PA}} + R_A = \\ = R_{EQ} + \frac{R_{PO}}{1+(j\omega)^{\alpha} C_{EDLO} R_{PO}} + \frac{R_{PA}}{1+(j\omega)^{\beta} C_{EDLA} R_{PA}} \tag{4}$$

where: $R_{EQ} = R_O + R_W + R_A$ is equivalent active resistance.

The approximation resulted in the following values of the parameters of the equivalent circuit of the EWH and anode, located in a conductive medium (water) are: $R_{EQ}$=77 Ohm, $R_{PO}$=2.9 Ohm, $C_{EDLO}$=0.07 F, $\alpha$=1.4, $R_{PA}$=1 kOhm, $C_{EDLA}$=0.05 F; $\beta$=0.45.

To find the impulse response of the equivalent circuit of the EWH-water-anode system (fig. 3), let us put down the impedance function (4) of the circuit in the operator form, replacing jω by s:

$$Z(s) = R_{EQ} + \frac{R_{CP}}{1+p^{\alpha} C_{EDLC} R_{CP}} + \frac{R_{AP}}{1+p^{\beta} C_{EDLA} R_{AP}} \tag{5}$$

Then the impulse response of the equivalent circuit of the EWH-water-anode system can be determined as the Laplace transform of the impedance function (5) of the circuit in the operator form:

$$z(t) = L\left[Z(s)\right] \tag{6}$$

$$z(t) = R_{EQ}\delta(t) + \frac{t^{\alpha-1}}{C_{EDLC}} E_{\alpha,\alpha}\left[-\frac{t^{\alpha}}{R_{CP}C_{EDLC}}\right] + \frac{t^{\beta-1}}{C_{EDLA}} E_{\alpha,\alpha}\left[-\frac{t^{\beta}}{R_{PA}C_{EDLA}}\right] \quad (7)$$

where: $E_{\alpha,\alpha}[\bullet]$ is Mittag-Leffler function.

The impulse response graph of the equivalent circuit of the EWH-water-anode system calculated using MATLAB R2015b is shown in fig. 6. To implement the above approach to modeling the equivalent circuit of the EWH-water-anode system by its impulse response in the Simulink/Simscape software environment, the model shown in fig.7 was developed.

The model, which is shown fig. 7, includes a Signal Builder, a controlled current source (I Source 2), a model subsystem (MODEL), an ammeter (Current Sensor), a voltmeter (Voltage Sensor), Simscape signal to signal converters Simulink and an oscilloscope that displays the current and voltage graphs of the model (Scope).

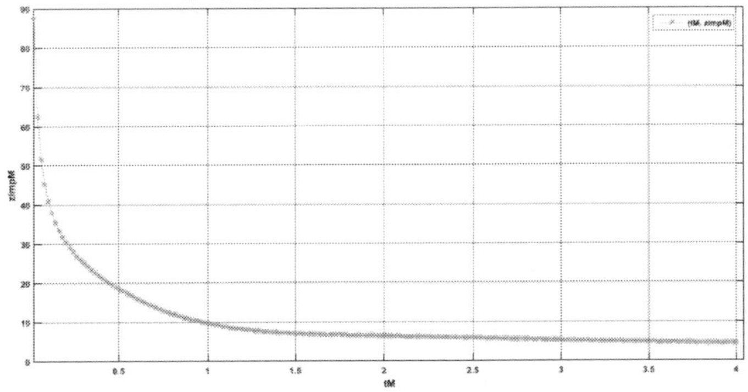


**Fig. 6.** The impulse response graph of the equivalent circuit of the EWH-water-anode system.

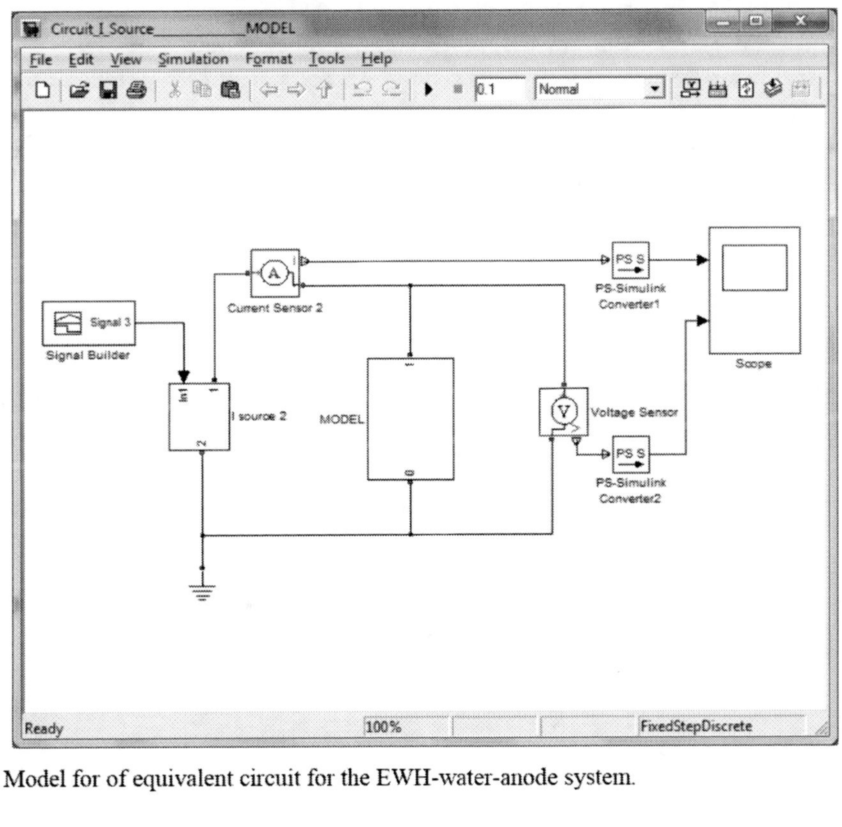


**Fig. 7.** Model for of equivalent circuit for the EWH-water-anode system.

## 5 Block diagram of a device for EWH corrosion protection potential diagnostics

By analyzing the process of diagnostics of the EWH corrosion protection potential (fig. 2) and the equivalent circuit of the object (cathode) and the anode, located in a conductive medium (fig. 3), the authors propose a block diagram of a device for EWH corrosion protection potential diagnostics which is shown in fig. 8.

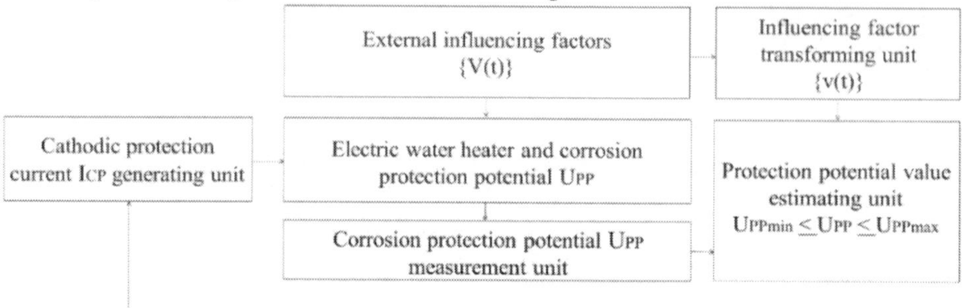


**Fig. 8.** General block diagram of a device for EWH corrosion protection potential diagnostics.

In the process of diagnosing the EWH corrosion protection potential, the cathodic protection current generating unit generates a cathodic protection current between the EWH body (cathode) and the anode $I_{CP}$.

In the process of the corrosion protection potential diagnostics, external influencing factors {V(t)} affect the EWH. The protection potential measuring unit measures $U_{PP}$, and the influencing factor transforming unit transforms them into various parameters of influencing factors {v(t)}.

The measured [9-12] protection potential $U_{PP}$ and the variety of influencing factor parameters {v(t)} are sent to the protection potential value estimating unit, which determines the assessment of Q (whether the protection potential $U_{PP}$ falls or not into the specified tolerance $U_{PP\min} \leq U_{PP} \leq U_{PP\max}$).

Next, two hypotheses are put forward.

1. The basic hypothesis $H_0$ (protection potential $U_{PP}$ falls into the specified tolerance $U_{PP\min} \leq U_{PP} \leq U_{PP\max}$).

$$H_0 : U_{PP} \in [U_{PP\min}, U_{PP\max}] \tag{8}$$

where: $U_{PP\min}$ and $U_{PP\max}$ are limits of permissible interval for the EWH corrosion protection potential $U_{PP}$.

2. Alternative hypothesis $H_1$ (protection potential $U_{PP}$ does not fall into the specified tolerance $U_{PP\min} \leq U_{PP} \leq U_{PP\max}$).

$$H_1 : U_{PP} \notin [U_{PP\min}, U_{PP\max}] \tag{9}$$

The instrumental diagnostic reliability of the EWH corrosion protection potential is defined by the expression, which includes the probabilities of errors of the first and second kind, α and β respectively.

The probabilities of errors of the first and second kind are defined according to the expressions:

$$\alpha_{U_{PP}} = \int_{U_{PP\min}}^{U_{PP\max}} f(U_{PP}) \left( \int_{-\infty}^{U_{PP\min}-U_{PP}} f(\delta_{U_{PP}}) d\delta_{U_{PP}} \right) dU_{PP} + \\ + \int_{U_{PP\min}}^{U_{PP\max}} f(U_{PP}) \left( \int_{U_{PP\max}-U_{PP}}^{\infty} f(\delta_{U_{PP}}) d\delta_{U_{PP}} \right) dU_{PP} \quad (10)$$

$$\beta_{U_{PP}} = \int_{-\infty}^{U_{PP\min}} f(U_{3П}) \left( \int_{U_{PP\min}-U_{3П}}^{U_{PP\max}-U_{3П}} f(\delta_{U_{PP}}) d\delta_{U_{PP}} \right) dU_{PP} + \\ + \int_{U_{PP\max}}^{\infty} f(U_{PP}) \left( \int_{U_{PP\min}-U_{PP}}^{U_{PP\max}-U_{PP}} f(\delta_{U_{PP}}) d\delta_{U_{PP}} \right) dU_{PP} \quad (11)$$

where: $U_{PP\min}$ and $U_{PP\max}$ are limits of permissible interval for the EWH corrosion protection potential $U_{PP}$;

$f(U_{PP})$ is density of probability for the EWH corrosion protection potential $U_{PP}$;

$f(\delta_{U_{PP}})$ is density of measurement error probability for the EWH corrosion protection potential $U_{PP}$.

Then the general reliability of the EWH corrosion protection potential diagnostics is defined by the expression

$$D_G = D_M \cdot D_I = D_M \cdot (1 - \alpha - \beta) \quad (12)$$

Assessment of the quality of the EWH corrosion protection potential diagnostics may also be calculated by the formula of the average diagnostics risk

$$\Psi \approx W_{1U_{PP}} \alpha + W_{2U_{PP}} \beta \quad (13)$$

where: $W_{1U_{PP}}$ and $W_{2U_{PP}}$ are losses due to errors of the 1st and 2nd kind.

## 6 Conclusions

By analyzing the equivalent circuit that simulates the process of diagnosing the EWH protection potential, we may conclude that traditional automatic cathodic protection devices measuring the cathodic protection potential $U_{PP}$ when the cathodic protection current $I_{CP}$ passes through, make a relative methodological error in measuring the cathodic protection potential $\delta_M$.

Methodological measurement error $\delta_M$ reduces the instrumental reliability of diagnostics of the cathode $D_I$ protection potential and, consequently, the general reliability of the EWH corrosion protection potential diagnostics $D_G$.

Methodological measurement error $\delta_M$ is conditioned by the voltage drops that occur due to cathodic protection current $I_{CP}$ passing through the active resistance of electrolyte, active resistance of the EWH body (cathode) and active resistance of connecting wires.

The value of the methodological measurement error $\delta_M$ may be described by the expression

$$\delta_M = \frac{U_C + U_{EC} + U_{CW}}{U_{PP}} = \frac{I_{CP}(R_C + R_{EC} + R_{CW})}{U_{PP}} \quad (14)$$

On analyzing the expression of the methodological measurement error $\delta_M$, we may conclude that an increase in the cathodic protection current $I_{CP}$ and active resistances $R_{EC}$,

$R_C$ and $R_{CW}$, leads to an increase in the methodological measurement error $\delta_M$ and a decrease in the instrumental and general reliability of the EWH protection potential diagnostics.

## References

1. W. Baeckmann. Handbuch des kathodischen Korrosionsschutzes: Theorie und Praxis der elektrochemischen Schutzverfahren / W. Baeckmann, W. Schwenk – Weinheim: Viley-VCH, 1999 – 530 p., (1999)
2. K.J. Kim, M.H. Lee and K.M. Moon, "Principle and Application of Cathodic Protection for Concrete structures", Korea Concrete Institute (KCI) journal, (2006)
3. S. Wong, W. Muneer, S. Nazir, A. Prieur, Designing, Operating, and Simulating Electric Water Heater Populations for the Smart Grid, Report No. 2013-136 (RP-TEC), Canmet ENERGY, Varennes Research Centre, Natural Resources Canada, October (2013)
4. D. C. Montgomery and G.C. Runger, Applied Statistics and Probability for Engineers. 2nd Edition, John Wiley & Sons, New York, (1999)
5. V. Martynyuk, O. Eromenko, J. Boiko and T. Kałaczyński. Diagnostics of supercapacitors. MATEC Web Conf., **182**, 01009, (2018)
6. V. V. Martynyuk Fractional Model of an Electrochemical Capacitor / V. V. Martynyuk, M. D. Ortigueira // Journal of Signal Processing. – Vol. 107, pp. 355-360, (2015)
7. V. Martynyuk, O. Vdovin, J. Boyko and N. Vlasenko "Super-high capacitor analyzer with compensation of common-mode error," in Proc. IMECO TC-4, pp. 340–343, (2001)
8. V. Martynyuk, D. Makaryshkin, and J. Boyko, "Frequency domain analysis for electrochemical supercapacitors", in 2007 Proc. 15th IMEKO Symposium on Novelties in Electrical Measurements and Instrumentation in Parallel with the 12th Workshop on ADC Modelling and Testing (IMEKO), Iasi, pp. 5., (2007)
9. S. Galla, A. Szewczyk, J. Smulko and P. Przygocki: Methods of Assessing Degradation of Supercapacitors by Using Various Measurement Techniques // Applied Sciences-Basel. -Vol. 9, iss. 11, s.1-11, (2019)
10. V. Martynyuk, D. Makaryshkin and J. Boyko, "Electrochemical supercapacitor time domain analysis by means of multi-channel measurement system", in 2007 Proc. 15th IMEKO Symposium on Novelties in Electrical Measurements and Instrument. in Parallel with the 12th Workshop on ADC Modelling and Testing, Iasi, Vol. I, pp. 207–211, (2007)
11. M. Łukasiewicz, T. Kałaczyński, J. Musiał, J. Shalapko, *Diagnostics of buggy vehicle transmission gearbox technical state based on modal vibrations,* Journal of Vibroengineering **16, 6**, (2014)
12. . Kałaczyński, M. Łukasiewicz, J. Musiał, R. Polasik, M. Szczutkowski, N. Dluhunovych, J. Wilczarska, T. Kasprowicz. *Analysis of the diagnostic potential research thermovision in the technical state of combustion engine injectors assessment,* INT CONF ENG MECHS , Engineering Mechanics,(2018)

# Assessment of friction pair elements condition based on changes in the geometric surface structure isotropicity degree

*Maciej* Matuszewski[1,*], *Małgorzata* Słomion[1], *Adam* Mazurkiewicz[1] and *Danil* Yu. Pimenov[2]

[1]University of Science and Technology in Bydgoszcz, Faculty of Mechanical Engineering, 85-796 Bydgoszcz, Poland
[2]South Ural State University, Department of Automated Mechanical Engineering, Lenin Prosp. 76, Chelyabinsk 454080, Russia

**Abstract.** This paper regards the possibility of using the degree of the surface isotropicity to assess the condition of the surface layer friction pair elements. The analysis is based on certain changes in operational and technological surface layer. The methods of constituting the surface layer have been characterized due to the features of geometric surface structure as well as the problem of assessing the condition of surface in terms of functional features and tribological characteristics.
The evaluation of the occurring changes in the surface layer have been conducted by performing tribological tests and for descript condition of surfaces the degree of surface isotropicity and weight loss of specimens have been adopted. Whereas the assessment of technological state of the surface layer has been carried out for the surface after its coating. Based on the analysis, it has been found that the use for assessment the condition of surface layer and the degree of the surface isotropicity are useful and it is an alternative way or supplement to classical adoption of a properly selected set of roughness parameters to evaluate the potential functional characteristics of friction couples surfaces.

## 1 Introduction

Each technical object has a set of functional qualities, such as: reliability, wear resistance, fatigue strength, resistance to movement, corrosion resistance, etc. Based on numerous studies, it may be concluded that these characteristics depend on the greatest extend of the condition of the surface layer, especially on the parameters characterizing the geometric surface structure [1-8]. Due to the fact that most these damages happen inside of the surface layer, the proprieties enlarging- resistance on destructive working extortions during machine and devices operation are given rught to the layer. Therefore, the surface layer is the subject of researches conducted in many scientific centers: domestic and foreign ones. It has been confirmed that the mechanism and intensity of wear process depend on features of the surface layer, established as a result of technological processes [9-12]. Affected by these factors (for example: mechanical, thermal, chemical and electrical) while the friction pair is working, the surface layer is changing. These changes usually cause a different loss degree of kinematic pairs ability to work to fulfill structurally accepted functions.

Shaping desirable features of the surface layer, especially the geometric surface structure, already in the manufacturing process is a very important problem due to the ability of kinematic pair elements to fulfil the technical object functions adopted in the construction process as long as possible. The surface layer produced in a technological process has a significant impact on functional characteristics, which determines the durability of the entire machine. It should be remembered that tribological processes should be always considered for a pair of elements- depending on the association with the intensity of the wear process and even its mechanism may differ. The characteristics of the surface layer are largely determined by stereometric parameters, which describe surface stereometric structure, more often called the geometric surface structure. They are most commonly used to describe changes in the surface layer. They are characterized by continuous changes during the manufacturing and operating process and thus describe well the current state of the surface layer. During the cooperation of elements in a friction pair, their surface is in the interaction and therefore the geometric surface structure has a significant impact on functional features and tribological characteristics [11, 13-18].

Taking into account the phases of the machine existence ("life"), two forms of surface layer and its elements could be distinguished: technological and operational. They occur as follows:

- in a phase of construction, design, machine elements manufacturing and their assembly (technological surface layer),
- in a phase of the machining operation (operational surface layer).

There is no ambiguous information in the literature how should be surface layer elements classify during the warehousing and transport phase, into the operational surface layer or technological surface layer. On the one hand, it is not yet affected by external operational forces, so there are no grounds to include it in the operational surface layer, on the other- the manufacturing process has been already completed- so it is no longer a technological surface layer.

In this work, the degree of surface isotropicity was assumed to assess the condition of friction pair elements.

## 2 Methods of the surface layer constitute

There are many methods for constituting the technological surface layer, as well as the number of different treatments and technological operations- the components of technological processes of machine elements. A very general division of the surface layer creating methods in which the criterion of division is the increase of material was proposed in [10, 19-22] studies. According to this classification, the methods of the surface layer constituting are divided into:

*Corresponding author: matus@utp.edu.pl

- weight- loss method- realized by reducing object dimensions, e.g. machining, erosion, blasting work,
- loss- free method- implemented without reducing the dimensions of the object, e.g. heat treatment, plastic working, implantation, PVD, CVD,
- incremental method- realized by increasing the dimensions of the object, e.g. electrolytic coating.

The examples of listed treatments in each of the cited groups present the fact that the increase is generally treated, i.e. it might be positive, negative or neutral. In addition, it may be mass or geometric. Within the division adopted in this way, all known techniques used in manufacturing processes shall be classified at least into one of the cited groups . It should be noted that the mass increase is not always accompanied by a change in the linear dimension and vice versa. Representative examples in this regard may be:

- implantation - weight gain without geometric dimension changing,
- burnishing - changing the geometric dimension without mass changing.

Based on the above considerations, it could be concluded that the technological surface layer is dynamic- its condition changes after each cut or operation. However, the functional characteristics of machine is determined by the condition of the surface layer elements after last operation, that is why this condition should be adopted as the technological surface layer of the finished element and then it does have a static character.

The weight-loss and loss-free treatment are the basic manufacturing techniques used to shaping elements of the technical devices.

The most widespread, because of its great universality and accuracy of shaping is the cutting process based on the model of removing the material allowance in a form of chips by the cutting tool edge. The total share of machining in the machine industry is currently estimated at about 50% and according to the forecasting by the International Institution for Production Engineering Research (CIRP) this share will be significant for a long time. It results in growing possibilities of machining applications and received accuracy [19].

Dynamically developing techniques for modifying the condition of the geometric surface structure of frictional cooperating elements at the production stage are coating application techniques by the physical or chemical methods. A thin- as low as a few micrometers- ceramics layer (most often) reduces the intensity of the wear process. One of the reasons for this phenomenon is the fact that applied coating has a significantly higher hardness than the substrate and thus more resistant to the effects of external operational forces. Application of the coating causes changes in the surface structure, which may also contribute to reduction of the intensity of the wear process.

Changes of the geometric surface structure condition are dependent on a coating method. Physical vapour deposition (PVD) and chemical vapour deposition (PVD) are basic methods for getting thin layers.

Conventional (i.e. not supported) techniques for producing nanocrystalline layers by CVD methods depends on the coating material setting from the gaseous phase as a consequence of chemical reactions occurring the substrate surface or in its vicinity. Temperature, chemical composition and conditions for flow of reagents through the retort are the parameters that affect this process.

The processes of coatings physical vapour deposition (PVD) consist of changing the physical state of the coating material based on vacuum techniques. There are two basic methods to change the physical material state from which the coating will be made:

- evaporation or sublimation,
- sputtering influenced by non-thermal physical excitations.

PVD coatings, like CVD, significantly increase wear resistance and thus durability. The application is also similar, i.e. mainly these are coatings applied to cutting tools and frictional cooperating surfaces.

## 3 Transformation of the surface layer

The features of technological surface layer, constituted as a result of the implementation of the assumed technological process, significantly depend on kind of operations and their parameters that occur in a manufacturing process. In turn, the course of wear depends on the features of the technological surface layer- its mechanism, intensity, effects and consequently the machines durability. Characteristics of the technological surface layer are changing after each operation and treatment undertaken, so it might be said that it is dynamic. These changes concern both, the inside of surface layer and its top and the whole process is called the transformation of the technological surface layer. However, the functional characteristics of machine is determined by the condition of the surface layer elements after last operation, that is why this condition should be adopted as the technological surface layer of the finished element and then it does have a static character.

The effect of kinematic excitations is mainly observed in the geometric surface structure. These forces are affected by kinematic machining parameters. The shape and placement of machining marks after each technological operation, i.e. the directionality of the geometric surface structure, depend on their value and relation between the movements of the tool and workpiece. Figure 1 shows examples of surface structures obtained by various finishing methods.

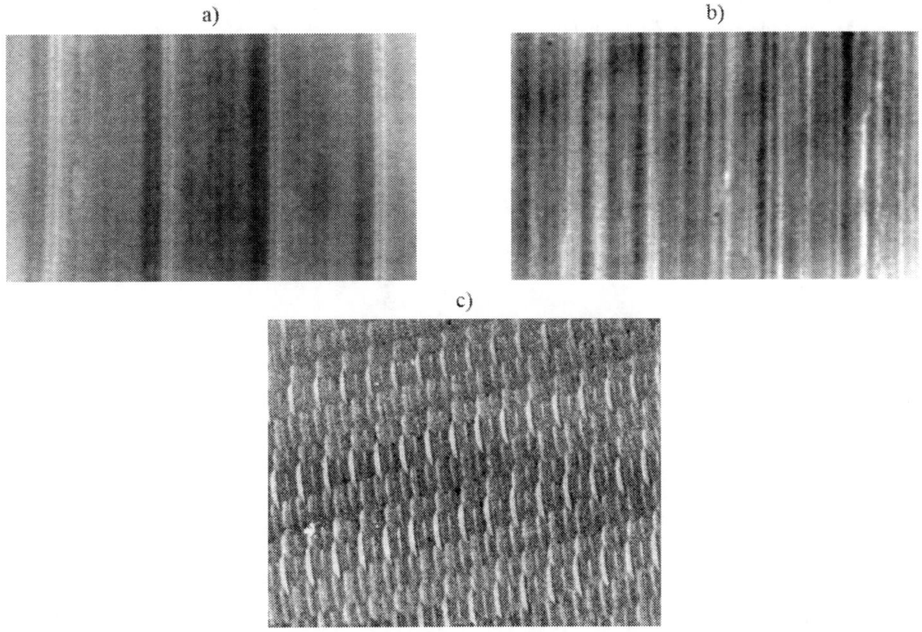


**Fig. 1.** Surface structures after finishing: a) turning, b) grinding, c) mill-turning.

The presented drawings show that the structures obtained by turning and grinding are anisotropic, while in the case of hybrid machining which is milling, the structure is clearly isotropic. As the operational characteristics of technological surface layer is adopted those which characterize the condition of surface layer after completion of the whole technological process. Transient parameters describing the state of surface layer are significant, only if they cause any reactions in following technological operations.

Since the moment of beginning of operation process of a technical object, right from the time when it begins to affect the object and its component elements the external operational forces, the technological surface layer is transformed into operational surface layer and the process of transformation begins. Similarly, like in the case of technological surface layer transformation, it is a dynamic process. The state of operational surface layer is changing during the whole phase of machines and devices operation. The common feature of transformation, both technological surface layer and operational surface layer, is their dynamic character. Nevertheless, in case of technological surface layer, with regard of the course of operation process, a really essential is the state at the end of production, however for operational surface layer essential is its current state.

For description of state and changes in the surface layer of machine elements, and so to delimitate also the functional characteristics of machines, the most often the different parameters of roughness are applied [1, 23-26]. Undoubtedly, they are important for both the splitting and concentrated contact of cooperating surfaces, however the inference of the condition of the surface layer only based on roughness parameters may not reflect the surface functional behaviour the specific external forces.

## 4 The issue of evaluating the status of the geometrical surface structure in the aspect of functional characteristics

One of the fundamental objectives of geometrical surface structure metrology in addition to control is to anticipate and optimize its functional features of the surface which are related to its shape. An important step in the analysis of the surface is to choose the appropriate parameters which lead to proper conclusions, whether the determinate geometric structure can provide the necessary functional characteristics of surface. Facilitating factor in choosing parameters for the geometric surface structure analysis is to consider the working conditions and resulting goals [11, 15, 18, 21, 26,27]. These conditions might be, for example:

- should the surface hold the lubricant?
- will the material be removed from the surface?
- will the surface tension occur?

It should be clear that to take an advantage of all parameters to access the geometric surface structure with respect to its functional characteristics is not realistic. On the other hand, the simplification is to attempt a geometric surface structure using only one parameter, which is unfortunately very common in industrial practice. The development of computer-aided, precision measuring and digital apparatus for surface analysis favors the creation of many parameters used to description the geometric surface structure. However, there is a tendency to harmonize and reduce the number of privileged parameters in national standards these days. Creating a universal set of parameters, equally useful for surfaces with various functional assignment is not possible in practice. Even with similar values of basic amplitude parameters, the other parameters may differ significantly. The surface assessment may be then ambiguous, because they shall have similar or the same roughness parameter value and different functional characteristics. Figure 2 shows an example comparison of Ra roughness parameter for different types of profiles.

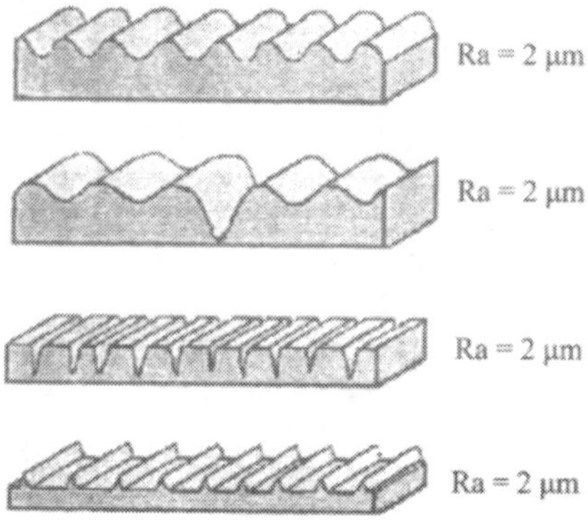


**Fig. 2.** Comparison of profiles with the same Ra value [27].

Despite the same value of the Ra roughness parameter, presented profiles will have different operational features. For each of analyzed surfaces should be decided separately which features of the geometric surface structure are important in terms of functional characteristics and which roughness parameters give the most appropriate information. No standards exist that stating in which cases particular parameters should be used. Only relevant recommendations can be found in the literature to selecting parameters for evaluation the functional characteristics.

In figure 3 the examples of roughness parameters comparison for different surfaces determining different functional characteristics are shown.

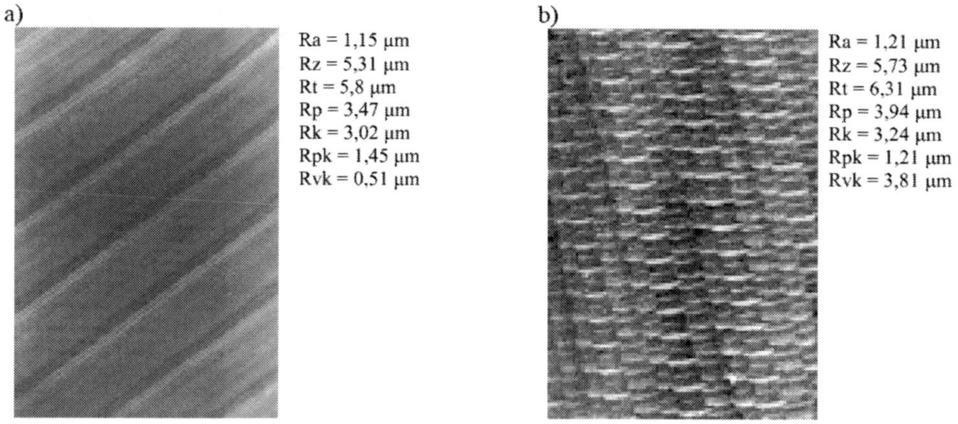


**Fig. 3.** View and parameters describing the geometrical surface structure of the machined surfaces: a) turning - for vC = 18.5 m · s-1 and f = 0.17 mm · rev-1 (aluminum), b) mill-turning - for vC = 2, 1 m · s-1 and f = 0.75 mm · rev-1 (steel)

The presented comparison shows that the amplitude parameters, which are commonly used to describe tribological characteristics, and which were the first parameters taken into account to assess the surface, do not quite accurately inform about the possible behavior of the surfaces in a kinematic pair. Only the analysis of the bearing capacity curve parameters differentiates frictional association behavior. A surface with a much higher Rvk value will be the surface with greater oil holding ability and at the same, because of the lower value of Rpk, although not so distinctly, it will be more wearability in initial phase of cooperation.

Selection of various roughness parameters for the assessment functional characteristics and to appoint tribological characteristics is carried out, i.a. based on researches, statistical relationships between them and availability of apparatus. As already shown, the selected parameters describing surface structure in the aspect of its functional characteristics cannot be always clearly classify. Surface characteristics based on roughness parameters are additionally difficult because of the parameter values variability at the different surface places which resulting from the machining process. For each parameter the range of its value is different, usually larger for longitudinal parameters than for height ones and in practice the possible spread can reach up to several percent. For these reasons, this selection will always be burdened with more or less subjectivity. In functional characteristics evaluation is always recommended to use a set of appropriately selected parameters, which is often complicated [12, 15, 18, 21, 24, 26]. Even in some studies concerning the functional characteristics assessment besides selected for evaluation parameters it is possible to find an information about processing method. Therefore, some beneficial solution is to use mathematical functions, which describe the surface in a frequency method, for delineate the geometric surface structure [22, 28]. This procedure seems to be free from the fault of subjective parameters selection model and with the dynamic development of computer techniques supporting engineering activities, determining these functions is not a complicated activity. In addition, frequency analysis allows the assessment of the geometric surface structure, independently from the type of treatment [22, 28]. The most reasonable and precise in terms of mathematical description, seems to be the method based on the analysis of surface autocorrelation function or spectral power density [22, 28]. In practice, one of these functions can be used for analysis, because the spectral power density function is a Furier transformation of the surface autocorrelation function. In general, this function is a measure of dependence between value of data in one location and its value in the other. Application of space-frequency functions is becoming more and more common not only to assess changes in the surface layer in terms of their functional characteristics. The frequency analysis is the source of information about the manufacturing process, e.g. as a carrier of information about condition of the cutting tool blade or for monitoring the technological process based on e.g. the value of cutting forces and their components [20, 21, 24, 25].

The graphic form of the autocorrelation function can be presented, e.g. in the form of so called rose of directions [18]. This rose determine main directions of surface formation. From the graph (Fig. 4.) can be inferred only qualitatively about the nature of surface which is shown in Figure 5. However, from the function and indirectly from the graph you can determine the quantitative parameter - the degree of surface isotropicity, that is, the degree of orientation - ordering - traces of machining or wear.

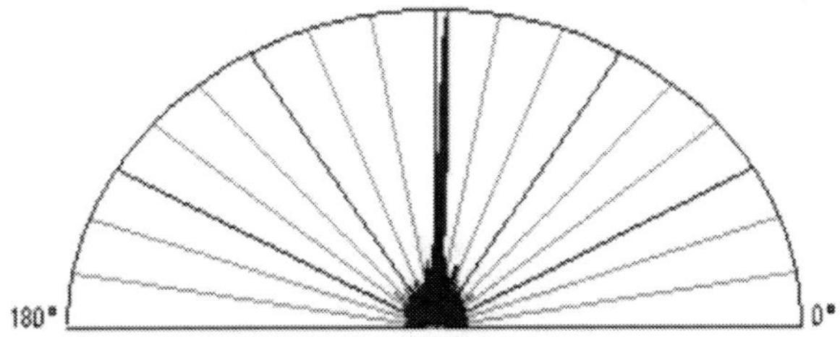


**Fig. 4.** Example of surface structures directions rose- degree of isotropicity: Iz = 8.01%.

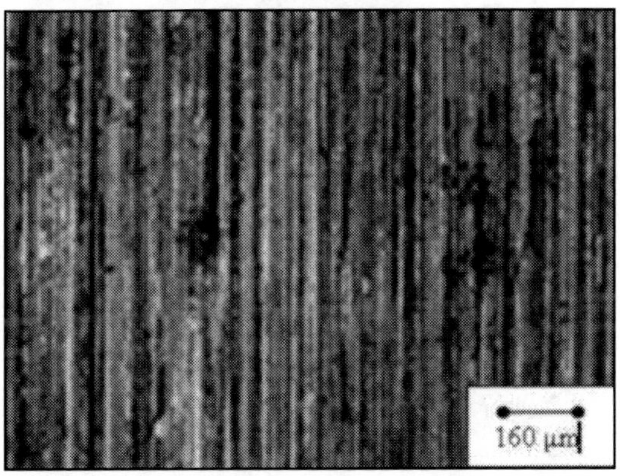


**Fig. 5.** Image of the geometric surface structure obtained by a confocal laser scanning microscopy.

## 5 Results

In order to verify the usefulness of using the degree of isotropicity for the quantitative assessment of occurring changes in the surface layer of cooperating surfaces, experimental tests have been carried out. They have been made in two stages. First, tribological tests were carried out, so the usefulness of applying the degree of isotropicity to assess changes in the operational state of the surface layer was analysed. For the quantitative assessment of changes, the degree of isotropicity and additional weight loss were adopted as a direct measure of wear. During surface cooperation, there is a continuous change in the shape of the geometric surface structure, so these tests are a good tool for verifying the suitability of using a degree of surface isotropicity to describe the shape of the structure, which in turn determines the surface usability characteristics. The weight loss and changes in the degree of isotropicity during the transformation of the surface layer reflect well the current state of the surface layer, and thus the current functional characteristics. The weight loss takes place on the whole surface, so it applies to changes in the entire area of cooperating surfaces. Structure isotropicity also applies to the entire area, it is not local. Conditions during tribological tests were mostly adopted as permanent. Specimens with counter-specimen cooperated in the lubricating medium, which was machine oil (L-AN 68), and the values of operational parameters were as follows: relative motion speed: 2,9 m/min (0,05 m/s), cooperation angle between characteristic directionality of specimens and counter-specimen was 0°. Specimens were made of 102Cr6 steel, whereas counter-specimen was made of X210Cr12 steel. Counter-specimen hardness (60 HRC) exceeded the hardness of the specimens so that changes took place primarily on the specimens surfaces. Hardness values of the specimens were: 30, 40 and 50 HRC. Specimens worked at 300, 450 and 600 N loads, which corresponded to the theoretical pressures in the contact zone: 1; 1.5 and 2 MPa. Measurement points were adopted for friction distance at which the observed changes were the most intense and amounted to: 100, 200, 300, 400, 500 and 600 m.

Figure 6 presents changes in the degree of isotropicity of the geometric surface structure of tested specimens and loss of their mass as a function of friction distance.

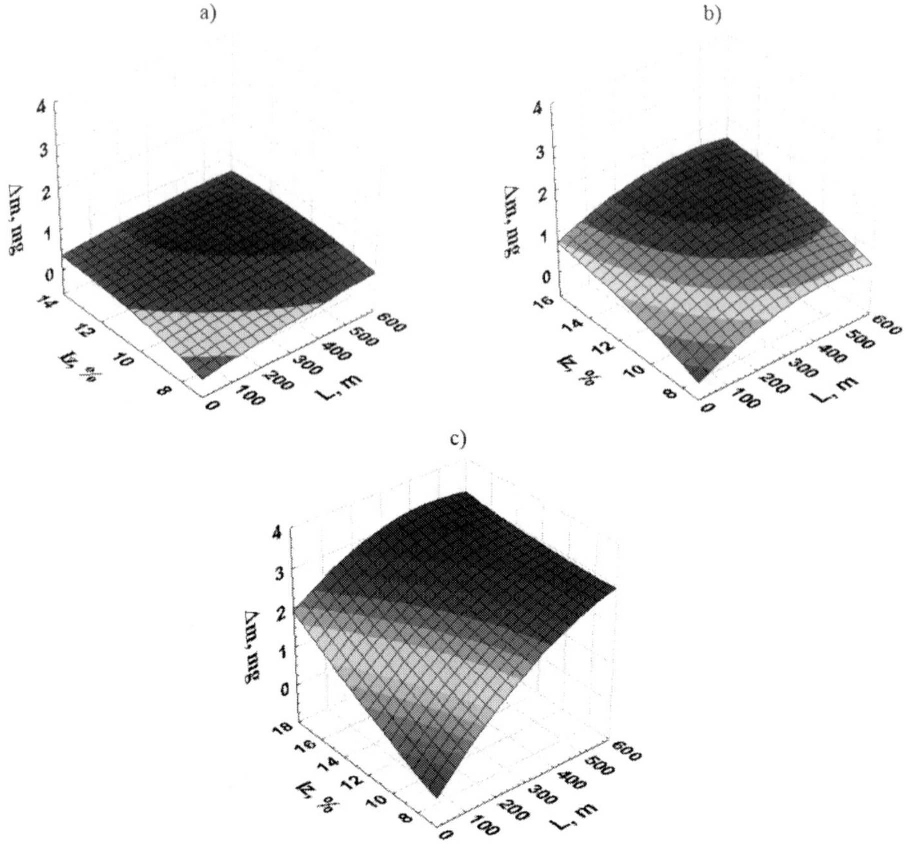


**Fig. 5.** Relations between mass loss of specimens *Δm*, isotropicity degree *Iz* and friction distance *L* for: a) $F_1$ = 300 N i $H_3$ = 50 HRC, b) $F_2$ = 450 N i $H_2$ = 40 HRC, c) $F_3$ = 600 N i $H_1$ = 30 HRC.

Based on the results presented in the drawings it shall be stated that the analysed factors and individual relations confirmed impact of the isotropicity degree on the surface layer transformation. The greatest changes are observed for external operational factors (F = 600 N and H = 30 HRC), which as expected, should cause the most intense changes, and the smallest- also as expected, for the external operational factors: F = 300 N and H = 50 HRC. With the least intense changes in the surface layer, also changes in isotropicity and mass are clearly noticable. Changing the degree of isotropicity is a change in the shape of the geometric surface structure. The greater the degree of isotropicity, the distribution of characteristic elements of structure is more random. They are "flattened" and the symmetry of this configuration increases (the degree of isotropicity increases). Single peaks and microroughness ridges are subject to partial or complete wear. Due to the configuration structure becomes more symmetrical, which can have a positive effect on the distribution of the lubricating medium and as a consequence on the resistance to motion.

Based on carried out studies it could be concluded that dependence of the operational surface layer transformation course from the degree of isotropicity, which is characterizing the entire potential contact is beneficial due to the designated tribological characteristics.

At the second stage of tests the usefulness of applying the degree of isotropicity to assess changes in the technological surface layer state by coating were analysed. Primary surfaces with a clear lack of structure orientation (isotropic) and surfaces with visible directionality of the geometric surface structure (anisotropic) were observed and compared with the state after coating. The topography of these surfaces is shown in Figures 7 and 8, and the values of measured parameters are summarized in Table 1. Isotropic surfaces of tool bit made of self-bonded carbide (fig. 7.), while anisotropy was characterized by ground surfaces of a twist drill blade made of 102Cr6 tool steel (fig. 8). In both analysed cases, the coating was made of titanium nitride (TiN).

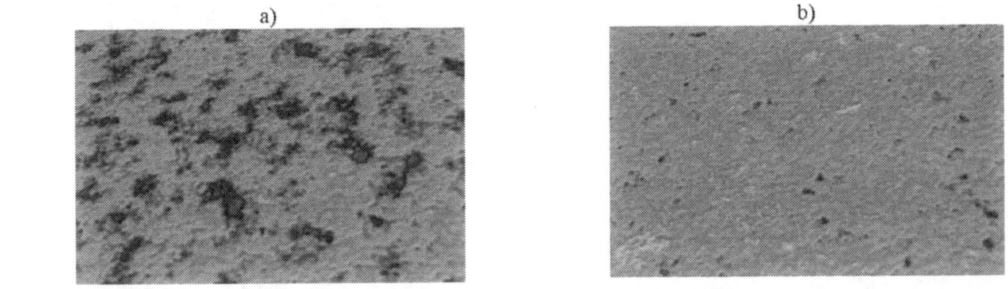

**Fig. 7.** Topography of self-bounded carbide surface: a) without coating, b) with TiN coating.

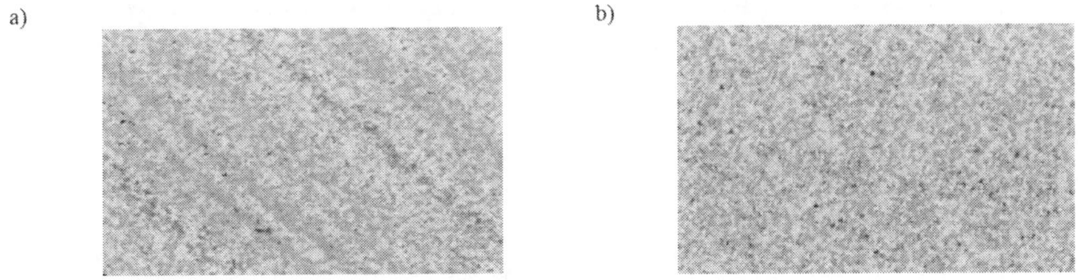

**Rys. 8.** Topography of the twist drill surface made of 102Cr6 tool steel: a) without coating, b) with TiN coating.

Table 1. Calculated values of the geometric surface structure parameters

| Measured (calculated) parameter | Substrate material | | | |
|---|---|---|---|---|
| | self-bounded carbide | | tool steel 102Cr6 | |
| | without coating | coated | without coating | coated |
| Ra, µm | 0,26 | 0,2 | 0,39 | 0,34 |
| Rq, µm | 0,33 | 0,26 | 0,45 | 0,39 |
| Rt, µm | 3,81 | 2,94 | 1,5 | 0,94 |
| Rpk, µm | 0,3 | 0,25 | 0,42 | 0,38 |
| Iz, % | 83,3 | 88,3 | 57,4 | 85,2 |

Based on the obtained results it shall be concluded that the surfaces with the coating are characterized by a lower value of roughness parameters than before coating and for all analysed parameters. In case of isotropicity, its varied change is observed after applying the ceramic coating to the tested materials. In case of surfaces with a clear lack of structure orientation (self-bounded carbide) changes are insignificant, whereas in case of surfaces with high anisotropicity (tool steel) changes are pronounced. In this case, the stereometric shape of surface after applying coating changed significantly, which reflects the change in the degree of isotropicity. On the other hand, roughness parameters do not characterize changes in shape, so they will not reflect well the changes in the functional characteristics of a friction pair consisting of such elements.

Based on the obtained tests it must be stated that determining the technological surface layer condition, which determines the potential functional characteristics, by using the degree of isotropicity may be beneficial due to the possibility of controlling the technological process with tribological criteria.

## 5 Summary

The usefulness of adopting the degree of surface isotropicity to assess the condition of the surface layer of cooperating friction pair elements and usefulness for assessing the condition of the surface layer at the stage of shaping desired characteristics of the geometric surface structure during the manufacturing process has been found on the basis of experimental research and analysis. Adopting the degree of isotropicity to assess the surface stereometric shape is an alternative or supplement method to the classic adoption of a properly selected set of roughness parameters, or individual parameters for the evaluation potential functional characteristics of the frictional surfaces couples. In addition, due to the possibility of subjective selection to assess the state of the surface structure roughness parameters, the use of the degree of surface isotropicity for this assessment may be a more advantageous solution, not loaded to the fault of the subjective selection model of parameters.

## References

1. J. H. Horng, J. F. Lin, K. Y. Li, *Effect of surface roughness on stell roller scuffing*. Wear, **184**, 203-212 (1995)
2. T. Kałaczyński, M. Łukasiewicz, J. Musiał, R. Polasik, M. Szczutkowski, N. Dluhunovych, J. Wilczarska, T. Kasprowicz: *Analysis of the diagnostic potential research thermovision in the technical state of combustion engine*

*injectors assessment*, 24th International Conference Engineering Mechanics 2018, Svratka, Czech Republic, Book of full texts, Institute of Theoretical and Appiled Mechanics of the Czech Academy of Sciences, Prague, 357-360 (2018)
3. J. Krawczyk, *Tribological properties of selected structural steels (Właściwości tribologiczne wybranych stali konstrukcyjnych)*. Tribologia, **4**, 223-233 (2010)
4. E. Mańka, M. Słomion, M. Matuszewski, *Constructional features of ropes in functional units of mining shaft hoist*. Acta Mechanica et Automatica, **12**, 1, 66-71 (2018)
5. M. Matuszewski, T. Mikołajczyk, D. Yu. Pimenov, M. Styp-Rekowski, *Influence of structure isotropy of machined surface on the wear process*. International Journal of Advanced Manufacturing Technology, **88**, 9, 2477-2483 (2017)
6. M. Matuszewski, M. Styp-Rekowski, *Influence of texture direction of kinematic pair elements surfaces on service operated layer transformation*. Tribologia **4**, 87-97 (2006)
7. J. Musiał, Ł. Muślewski, M. Pająk, A. Grządziela, *Analysis of vibration time histories in the time domain for propulsion systems of minesweepers*. Journal of Vibroengineering **17**, 3, 1309-1316 (2015)
8. M. Pająk, Ł. Muślewski, M. Pająk, B. Landowski, A. Grządziela, *Fuzzy identification of the reliability state of the mine detecting ship propulsion system*. Polish Maritime Research, **1** (101), 26, 55-64 (2019)
9. T. Burakowski, *The importance of surface engineering in tribology* (*Znaczenie inżynierii powierzchni w tribologii*). Tribologia, **4**, 1097-1111 (2002)
10. T. Burakowski, T. Wierzchoń, *Metal surface engineering* (*Inżynieria powierzchni metali*), (Publ. WNT, Warszawa, 1995)
11. R. Górecka, Z. Polański, *Metrology of the surface layer* (*Metrologia warstwy wierzchniej*), (Publ. WNT, Warszawa 1983)
12. I. M. Hutchings, *Tribology. Friction and wear of engineering materials*, (Elsevier, Cambridge, 1992)
13. A. Bustilo, D. Yu. Pimenov, M. Matuszewski, T. Mikołajczyk, *Using artificial intelligence models for the prediction of surface wear based on surface isotropy levels*. Robotics and Computer Integrated Manufacturing, **53**, 5, 215-227 (2018)
14. M. Łukasiewicz, T. Kałaczyński, J. Musiał, J. Shalapko, *Diagnostics of buggy vehicle transmission gearbox technical state based on modal vibrations*. Journal of Vibroengineering, **16**, 6, 3137-3145 (2014)
15. G. W. Stachowiak, A. W. Batchelor, *Engineering tribology*, (Butterworth-Heinemann, Great Britain, 2001)
16. M. Styp-Rekowski, *Geometrical constructional features of special rolling bearings against their exploitational properties*, Proceedings of IV$^{th}$ Symposium INTERTRIBO '90, **C**, 93-96 (1990)
17. M. Styp-Rekowski, E. Mańka, M. Matuszewski, M. Madej, D. Ozimina, *Tribological problems in shaft hoist ropes wear process*. Industrial Lubrication and Tribology, **67**, 1, 47-51 (2015)
18. P. Pawlus, *Surface topography: measurement, analysis, impact* (*Topografia powierzchni: pomiar, analiza, oddziaływanie*), (Publ. Rzeszow University of Technology, Rzeszów, 2006)
19. W. Grzesik, *Basics of cutting metal materials (Podstawy skrawania materiałów metalowych)*, (Publ. WNT, Warszawa, 1998)
20. W. Lubimow, K. E. Oczoś, *Selected issues of shaping surface irregularities in machining processes* (*Wybrane zagadnienia kształtowania nierówności powierzchni w procesach obróbkowych*). Mechanik, **3**, 81-84 (1997)
21. B. Nowicki B., *Geometric structure: surface roughness and waviness* (*Struktura geometryczna: chropowatość i falistość powierzchni*), (Publ. WNT, Warszawa, 1991)
22. K. E. Oczoś, W. Lubimow, *Surface geometric structure* (*Struktura geometryczna powierzchni*), (Publ. Rzeszow University of Technology, Rzeszów, 2003)
23. J. Raja, *Field testing of machine tool diagnostic techniques using surface metrology*. Annals of CIRP, **32**, 503-506 (1983)
24. B. G. Rosen, R. Ohlsson, J. Westberg, 1995. *Interactive surface modelling, an implementation of an expert system for specification of surface roughness and topography*. International Journal of Machine Tools and Manufacture, **35**, 317-324 (1995)
25. D. J. Whitehouse, *Process and quality control using surface finish*. Proceedings 6$^{th}$ Symposium IMEKO Metrology for quality control in production, Vienna, 699-710 (1998)
26. J. Williams, *Engineering tribology*, (Cambridge University Press, Cambridge, 2005)
27. M. Wieczorowski, A. Cellary, J. Chajda, *A guide to measuring surface roughness, or roughness and more (Przewodnik po pomiarach nierówności powierzchni czyli o chropowatości i nie tylko)*. (Publ. Poznan University of Technology, Poznań, 2003)
28. K. E. Oczoś, W. Lubimow, *Determinism and randomness of surface geometric structures* (*Determinowość i losowość struktur geometrycznych powierzchni (SGP)*). Pomiary Automatyka Kontrolna, **10**, 2002, 4-6 (2002)

# Method of assessment of the transport system functioning with the application of the simulation model of the operating process

*Klaudiusz* Migawa[1,*], *Agnieszka* Sołtysiak[1] and *Magda* Czyżewska[1]

[1]University of Science and Technology, Faculty of Mechanical Engineering, 85-796 Bydgoszcz, Poland

**Abstract.** In the article, there is presented the description of the algorithm of the operating process of the simulation model characteristics' determination realized in the transport system. The simulation model of the means of transport's operating process has been developed based on the that process's mathematical model (semi-Markov model). Due of the assumed principles, the constructed model can be applied for the purposes of assessment and control of the operating process because of criterion of different types. Analysis of performance's assessment of the operating means of transport is one of the ways of the possibilities of correct realization of the assigned transportation tasks in the transport systems. The simulation model of the operating process presented in the work, makes it possible to assess performance of the transport system's operation from the point of view of the selected assessment criteria: availability of the means of transport and the unit income generated in the process's modelling conditions. Based on the results of the tests conducted in the real system of the means of transport operating system, the model's input data have been developed and simulation experiments have been conducted. As an effect, there have been determined the exemplary values of the analyzed characteristics of the transport system's operation performance analyses.

## 1 Introduction

One of the ways making it possible to assess effectively and to control the quality of the complex technical objects in operating systems, is application of the mathematical and simulation models for describing and analysis purposes of the operating processes performed in these systems. Models of that type, may be applied for assessment and the quality's control of the complex technical processes from the point of view of different assessment criteria, such as costs, incomes, reliability, availability, safety, productivity [7, 11]. The issues concerning the operating process's steering performed in the complex technical systems depending on the selected assessment criterion, has been presented in many studies [1, 2, 3, 9, 12]. Depending on the type of selected research problems, appropriate mathematical models and methods of determining the optimum or quasi-optimum solutions are applied. Due to a considerable complexity of the processes realized in real systems of the technical objects operating, there occurs the necessity of applying appropriate methods and tools, including the stochastic models [1, 4, 5], and computer simulation programmes, ensuring effective performance of the researches with reference to the models of the considered operating processes, and analysis of the obtained results [6, 8, 10, 13, 14]. These models ensure the possibility of assessing and steering the operating performance of the complex technical systems from the point of view of selected criteria.

* Corresponding author: klaudiusz.migawa@utp.edu.pl

In the study there are presented the results of operating performance of a selected, real system of technical objects operating, obtained based on the researches of the developed, simulation operating process realized in the considered system. The operating process's simulation model presented in the study, makes it possible to assess performance of the transport system's operation on account of two assessment criteria: availability of the means of transport and the unit income generated as a result of the operating process's realization. The simulation model applied in the study, has been developed in order to ensure a quicker and easier analysis and possibility to consider different variants of the modelled operating processes (for ex. by the change of the values of the modelled process's parameters or the number of considered technical objects). The constructed operating process's simulation model has been developed based on the that process's mathematical model (semi-Markov model). The mathematical operating process' model, has been constructed based on the analysis of the operating states and events' spaces (operating process's occurrence model) concerning the technical objects (means of transport) operating in the analyzed real system.

## 2 Model of the means of transport operating process

### 2.1 Occurrence model of the operating process

Below there is presented the occurrence model of the analyzed means of transport's operating process. As a result of the identification of the examined means of transport's operating process, there have been determined the essential operating states of that process, and possible transitions between the singled-out states. On that basis, the graph of the operating processes states' changes presented in figure 1, has been constructed. The following operating states of the technical object have been singled out in the analyzed model: 1 – stoppage at the depot yard, 2 – carrying out of transportation task, 3 – stoppage at route caused by damage, 4 – stoppage caused by a collision, 5 – intervention following the collision, 6 – repair following damage or collision, 7 – preventive diagnosis, 8 – preventive repair, 9 – supply, 10 – servicing (on the day of use, periodical, seasonal).

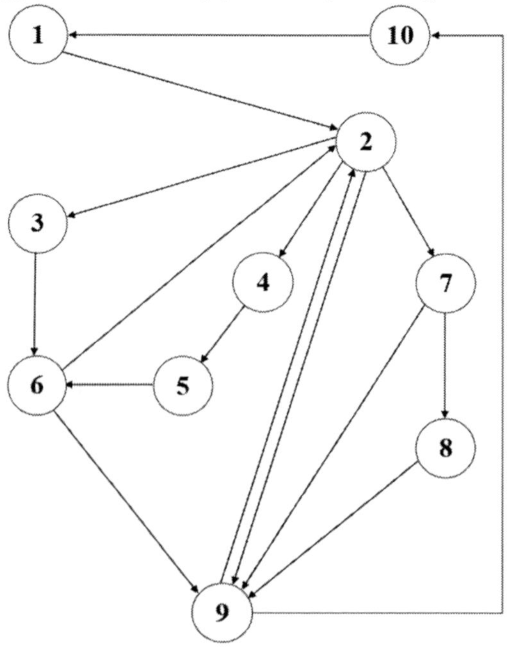


**Fig. 1.** Directed graph of the means of transport operating process's representing.

## 2.2 Mathematical model of the operating process

In the study, the mathematical model of the means of transport's operating process, has been built assuming, that the model of that process there is the uniform semi-Markov $X(t)$ process. The semi-Markov process $X(t)$ is the process, in which the times between the changes of subsequent process's states have arbitrary probability distributions and transition to the next state depends only on the current process's state. Applying the semi-Markov processes to mathematical operating processes' modeling, the following assumptions have been applied:

- modelled process has a definite number of states $i = 1, 2, \ldots, 10$,
- if a technical object at the moment $t$ is in the state $i$, then $X(t) = i$,
- random process $X(t)$, being the mathematical model of the operating process, is the uniform process,
- at the time $t = 0$, the process is in the state 1 (state 1 is the initial state), that means $P\{X(0) = 1\} = 1$.

The uniform semi-Markov process is determined unequivocally when there has been given the process's initial distribution and its kernel. From the accepted assumptions and based on the directed graph presented in figure 1, the initial distribution has the form of

$$p_i(0) = \begin{cases} 1, & gdy\ i = 1 \\ 0, & gdy\ i \neq 1 \end{cases}, \quad i = 1,2,\ldots 10 \tag{1}$$

where

$$p_i(0) = P\{X(0) = i\}, \quad i = 1,2,\ldots 10 \tag{2}$$

while the process's kernel $Q(t)$ has the form of

$$Q(t) = \begin{vmatrix} 0 & Q_{1,2}(t) & 0 & 0 & 0 & 0 & 0 & 0 & 0 & 0 \\ 0 & 0 & Q_{2,3}(t) & Q_{2,4}(t) & 0 & 0 & Q_{2,7}(t) & 0 & Q_{2,9}(t) & 0 \\ 0 & 0 & 0 & 0 & 0 & Q_{3,6}(t) & 0 & 0 & 0 & 0 \\ 0 & 0 & 0 & 0 & Q_{4,5}(t) & 0 & 0 & 0 & 0 & 0 \\ 0 & 0 & 0 & 0 & 0 & Q_{5,6}(t) & 0 & 0 & 0 & 0 \\ 0 & Q_{6,2}(t) & 0 & 0 & 0 & 0 & 0 & 0 & Q_{6,9}(t) & 0 \\ 0 & 0 & 0 & 0 & 0 & 0 & 0 & Q_{7,8}(t) & Q_{7,9}(t) & 0 \\ 0 & 0 & 0 & 0 & 0 & 0 & 0 & 0 & Q_{8,9}(t) & 0 \\ 0 & Q_{9,2}(t) & 0 & 0 & 0 & 0 & 0 & 0 & 0 & Q_{9,10}(t) \\ Q_{10,1}(t) & 0 & 0 & 0 & 0 & 0 & 0 & 0 & 0 & 0 \end{vmatrix} \tag{3}$$

where:

$$Q_{i,j}(t) = P\{X(\tau_{n+1}) = j, \tau_{n+1} - \tau_n \leq t | X(\tau_n) = i\} \tag{4}$$

$$Q_{i,j}(t) = p_{ij} \cdot F_{ij}(t) \tag{5}$$

$$p_{ij}(t) = P\{X(t) = j | X(0) = i\} \tag{6}$$

is the conditional probability of transition from the state $i$ to the state $j$

$$F_{ij}(t) = P\{\tau_{n+1} - \tau_n \leq t | X(\tau_n) = i, X(\tau_{n+1}) = j\} \tag{7}$$

is the distribution function of a random variable determining the duration time of the state $i$, on condition, that the next state there shall be state $j$.

The limiting probabilities $p_i^*$ of staying in the states of the considered process $X(t)$, have been determined based on the limiting theorem for the semi-Markov processes [1]

$$p_i^* = \frac{\pi_i \cdot \overline{\Theta}_i}{\sum_{i \in S} \pi_i \cdot \overline{\Theta}_i} \tag{8}$$

where:

$\overline{\Theta}_i$ – average unconditional duration times of the states of the process $X(t)$,
$\pi_i$ – probability of stationary distribution of the inserted Markov chain, which meets the system of linear equations

$$\sum_{i \in S} \pi_i \cdot p_{ij} = \pi_j, \quad j \in S, \quad \sum_{i \in S} \pi_i = 1 \tag{9}$$

where
$p_{ij}$ – conditional probabilities of transition from state $i$ to state $j$, according to the dependence

$$p_{ij} = \lim_{t \to \infty} p_{ij}(t) \tag{10}$$

Considering the above discussions, the following describing formulas have been determined:
– value of the technical object availability

$$A = \sum_{i \in S_A} p_i^* = \frac{\sum_{i \in S_A} \pi_i \cdot \overline{\Theta}_i}{\sum_{i \in S} \pi_i \cdot \overline{\Theta}_i} \tag{11}$$

where
$S_A \subset S = \{1,2,\ldots,10\}$ – set of availability states of the semi-Markov process model $X(t)$,
– value of the average unit income

$$C = \sum_{i \in S} p_i^* \cdot c_i = \frac{\sum_{i \in S} \pi_i \cdot \overline{\Theta}_i \cdot c_i}{\sum_{i \in S} \pi_i \cdot \overline{\Theta}_i} \tag{12}$$

where
$c_i$ – unit costs generated in the states of the process $X(t)$.

## 3 Simulation model of the performance assessment

The developed simulation model of the operating process, makes it possible to asses performance of the transport system's operation on acount of two assessment criteria: availability of the means of transport and the unit income generated as a result of the operating process's realization. The simulation programme, constituting the computer realization of the model of the means of transport operating model, has been constructed in just the same manner, making it possible to realize the simulation experiments with the use of the input data sets prepared based on operating data obtained from the researches concerning the real system of the means of transport's operating. The developed model of the operating simulation's programme, allows conducting simulation experiments for different numbers of operating occurrences (changes of the process's states), simulation time intervals, both in case of a single technical object as well as of a group of technical objects. In the simulation programme, subsequent realizations of the duration times of the operating process's model states, are determined with the use of the pseudorandom numbers' generating algorithms, generating the values from exponential, gamma, normal, log-normal and Weibull distributions. The structure of the simulation programme has been developed in such a way that to make it possible for the simulation experiment to represent in the course of realization the real set of technical objects (means of transport), and succession of occurrences finding fulfillment in the studies on the real transportation system, concerning individual technical objects. The developed simulation programme of the operating process' model was written by: R Core Team (2018). R: A language and environment for statistical computing. R Foundation for Statistical Computing, Vienna, Austria. URL http://www.R-project.org/. In figure 2, there is presented the block diagram of operation of the process of the technical objects operating simulation model's programme.

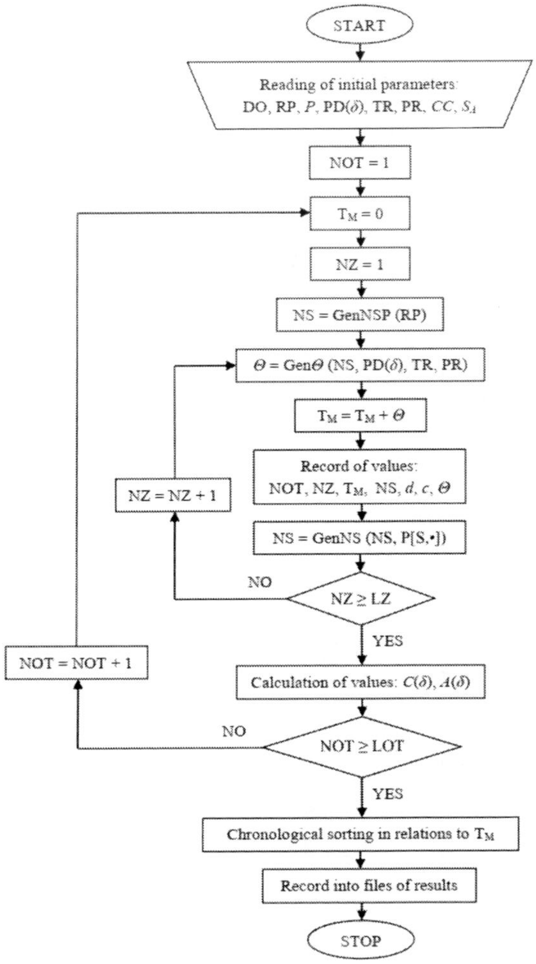


**Fig. 2.** Block diagram of the operation of the process of the technical objects operating simulation model's programme.

For the purposes of the simulation experiment's realization, it is necessary to prepare the set of the determined input data:

a) DO (general data characterizing the simulation experiment):
NOT – number of technical object in the simulation experiment,
NZ – number of occurrences for a technical object in the simulation experiment,
$m$ – number of states of the process, determined on the basis of the set of states of the modelled operating process $S = \{1,2,\ldots m\}$
ND – number of decisions possible to be applied in individual states of the modelled operating process,
N$\delta$ – number of strategies $\delta$ possible to be applied in the simulation experiment;
The strategy $\delta$ shall mean the sequence, the words of which are vectors, made of decisions $d_i^{(k)}(t_n)$ taken in individual states of the process $X(t)$, in subsequent moments $t_n$ of these states' changes

$$\delta = \left\{ \left[ d_1^{(k)}(t_n), d_2^{(k)}(t_n), \ldots d_m^{(k)}(t_n) \right] : \; n = 0,1,2,\ldots \right\} \tag{13}$$

In case when decisions taken in subsequent states of the process shall not depend on the moment $t_n$, in which they are taken, that is $d_i^{(k)}(t_n) = d_i^{(k)}$, then the strategy $\delta$ is called the stationary strategy. Then, the formula (13) shall assume the form

$$\delta = \left[ d_1^{(k)}, d_2^{(k)}, \ldots d_m^{(k)} \right] \tag{14}$$

b) RP (process initial distribution):
stochastic vector of the number of constituents equal to the number of the states of modelled operating process, determining distribution of staying in the states of the modelled operating process at the beginning of the simulation experiment;

c) *P* (probability matrix of the process's states changes):
square matrix of the degree equal to the number of the states of the modelled operating process, in which each line is the stochastic vector determining distribution of the process's states changes, on condition that the current state of the process shall be determined by that line's number;

d) PD (probability matrix of the decision's selection in the states of the process):
rectangular matrix of the number of lines equal to the number of states of the modelled operating process and the number of columns equal to the number of decisions possible to be taken in each state of the modelled process, composed of stochastic lines determining probability distributions of a given decision's selection at the moment of a technical objects entering a given state of a modelled process;
In case of applying of a deterministic strategy $\delta$, values of the matrix PD($\delta$) elements determine the stationary vector, the components of which are the numerical codes defining the deterministic decisions applied in individual states of the modelled operating process;

e) TR, PR (types and parameters of distributions items of remaining in the process states):
the developed programme includes the functions generating the values from the exponential, gamma, normal, log-normal and Weibull distributions on preset parameters;

f) *CC* (unit incomes matrix in the process states):
rectangular matrix of the number of lines equal to the number of states of the modelled operating process, and the number of columns equal to the number of decisions possible to be taken in each state of the modelled process, in which, for each state of the process and for each decision possible to be taken in a given state (at the meeting point of a line and a column), there is determined the unit income's value (per time unit) generated, when a technical object stays in a given state (determined by a line's number), while taking a decision (determined with the column's number);

g) $S_A$ (set of the process's availability states):
set $S_A$ of the numerical codes ascribed to the availability states of the operating process's model, determined based on the set $S$ of numerical codes ascribed to the operating process's model states ( $S_A \subset S = \{1,2,\ldots,m\}$ ).

Additionally, in figure 2, there have been used the following symbols: GenNSP – generating of the model's initial state number, GenNS – generating of the model's current state number, Gen$\Theta$ – generating of the value of time of staying in the current model's state.

In the course of the simulation experiment's conducting, at each moment in which change of the modelled operating process occurs (for a considered technical object), the following data are recorded in the results' file:

– number of the technical object NOT,

- number of the current occurrence NZ,
- moment of the current occurrence (current time of the model $T_M$),
- number of the current state of the model NS,
- current decision $d$,
- value of the unit income c concerning remaining of the technical object in the current state,
- generated value of the time of remaining in the current state $\Theta$.

Then, for the data set generated in the course of the simulation experiment's conducting, the values of characteristics of the transport system's operation performance assessment (for the applied strategy $\delta$) are determined:

- result determining the value of the technical objects availability

$$A(\delta)=\frac{\sum_{i=1}^{Z}\Theta_i(S_A)}{\sum_{i=1}^{Z}\Theta_i} \tag{15}$$

where:

$\Theta_i(S_A)$ – $i$-th realization of the time of remaining in the states of the modelled operating process, belonging to the availability states' sets $S_A \subset S=\{1,2,\ldots 10\}$,

$\Theta_i$ – $i$-th realization of the time of remaining in the states of the modelled operating process $S=\{1,2,\ldots 10\}$,

$Z = \text{NOT} \cdot \text{NZ}$ – number of occurences (changes of the model's states), for a determined number of technical objects;

- result determining the value of the average unit income

$$C(\delta)=\frac{\sum_{i=1}^{Z}c_i\cdot\Theta_i}{\sum_{i=1}^{Z}\Theta_i} \tag{16}$$

where:

$c_i$ – $i$-th realization of the unit income concerning remaining in the operating process's modeling states $S=\{1,2,\ldots 10\}$.

## 4 Exemplary results of the simulation model's researches

The presented example has been developed based on the operating data obtained from the researches concerning the real operating system of the means of transport (municipal transport services' buses). In the studied transport system, there are used 160 municipal transport service buses, and the service and repair processes are conducted at the technical back-up facilities' stations and by the technical service units. A group of 33 vehicles aged from 2 to 4 years has been selected for research purposes. For the considered model of the means of transport operating process, there have been determined the values of the simulation model's input parameters, and also based on operating data, the values of the elements of the transition probability matrices $P$, possible decisions taken in decisive process's states (table 1), and the mean values of the duration times and the mean unit incomes generated in the process's states (table 2) have been determined.

$$P = \begin{vmatrix} 0 & 1 & 0 & 0 & 0 & 0 & 0 & 0 & 0 & 0 \\ 0 & 0 & 0.2320 & 0.0030 & 0 & 0 & 0.4988 & 0 & 0.2661 & 0 \\ 0 & 0 & 0 & 0 & 0 & 1 & 0 & 0 & 0 & 0 \\ 0 & 0 & 0 & 0 & 1 & 0 & 0 & 0 & 0 & 0 \\ 0 & 0 & 0 & 0 & 0 & 1 & 0 & 0 & 0 & 0 \\ 0 & 0.4702 & 0 & 0 & 0 & 0 & 0 & 0 & 0.5298 & 0 \\ 0 & 0 & 0 & 0 & 0 & 0 & 0 & 0.0799 & 0.9201 & 0 \\ 0 & 0 & 0 & 0 & 0 & 0 & 0 & 0 & 1 & 0 \\ 0 & 0.2242 & 0 & 0 & 0 & 0 & 0 & 0 & 0 & 0.7758 \\ 1 & 0 & 0 & 0 & 0 & 0 & 0 & 0 & 0 & 0 \end{vmatrix}$$

**Table 1.** Decisions at decision-making states of the analysed process.

| Process state | Decision "0" | Decision "1" |
|---|---|---|
| 2 | Route with "light" conditions for completing the transport task | Route with "heavy" conditions for completing the transport task |
| 5 | Post-collision intervention – "intensive" | Post-collision intervention – "normal" |
| 6 | Reconditioning following damage or collision – extended scope | Reconditioning following damage or collision – basic scope |
| 7 | Preventive diagnosis – extended scope | Preventive diagnosis – basic scope |
| 8 | Preventive reconditioning – extended scope | Preventive reconditioning – basic scope |
| 9 | Supplying - intensive | Supplying - normal |
| 10 | Handling - intensive | Handling - normal |

**Table 2.** Mean lengths of time and unit revenues at the states of the analyzed process depending on decision.

| Process state | $\Theta_i^{(0)}$ [h] | $\Theta_i^{(1)}$ [h] | $c_i^{(0)}$ [PLN/h] | $c_i^{(1)}$ [PLN/h] |
|---|---|---|---|---|
| 1 | 5.659 | 5.659 | - 6.77 | - 6.77 |
| 2 | 8.852 | 7.967 | 28.44 | 33.56 |
| 3 | 0.514 | 0.514 | - 69.08 | - 69.08 |
| 4 | 0.873 | 0.873 | - 162.80 | - 162.80 |
| 5 | 1.712 | 2.372 | - 304.63 | - 234.11 |
| 6 | 2.633 | 3.342 | - 144.08 | - 118.21 |
| 7 | 0.441 | 0.677 | - 82.71 | - 54.98 |
| 8 | 1.665 | 2.140 | - 124.79 | - 96.89 |
| 9 | 0.096 | 0.152 | - 23.09 | - 17.46 |
| 10 | 0.354 | 0.582 | - 221.74 | - 164.72 |

Below there are presented the exemplary results of the simulation model's tests. The realized experiments consisted in 30 tuple simulation of the operating process for each considered strategy $\delta$. As a result of the simulation experiments' realization, there were obtained the sets of 30 values of the technical objects $A(\delta)$ availability and 30 values of the unit income $C(\delta)$. In the presented model, there have been distinguished the following technical objects availability states: 1, 2.

In figures 3 and 4 there are presented the exemplary results of simulation experiments for the selected strategies of the technical objects operating process's steering $\delta$, in such a manner, as the determined technical objects availability $A(\delta)$ be at least equal to the availability $A_r(\delta)$ required for correct realization of the transportation tasks (for $A_r(\delta) = 0.87$, 0.88, 0.89, 0.90). However, in the table 3 and in figures 5 and 6 there are presented the mean values, the values of standard deviation and the values of position statistics (minimum, 1 quartile, median, 3 quartile, maximum), determined for the considered characteristics of the technical objects performance assessment and for the selected strategies $\delta$.

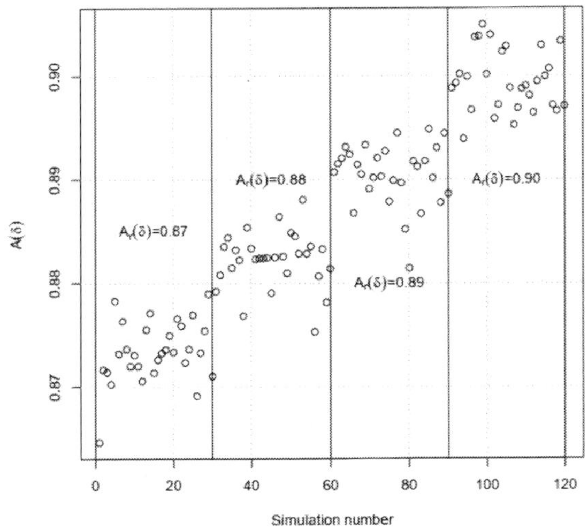


**Fig. 3.** Values of the technical objects availability determined based on the simulation experiments.

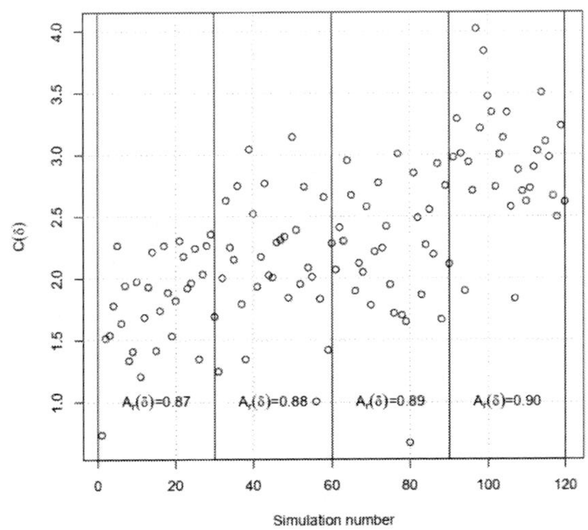


**Fig. 4.** Values of the unit income determined based on simulation experiments.

**Table 3.** Values of statistics determined for the technical objects availability and unit incomes in [PLN/h].

| | $A_r(\delta) = 0.87$ | | $A_r(\delta) = 0.88$ | | $A_r(\delta) = 0.89$ | | $A_r(\delta) = 0.90$ | |
|---|---|---|---|---|---|---|---|---|
| | $A(\delta)$ | $C(\delta)$ | $A(\delta)$ | $C(\delta)$ | $A(\delta)$ | $C(\delta)$ | $A(\delta)$ | $C(\delta)$ |
| Medium | 0.8734 | 1.8034 | 0.8822 | 2.1660 | 0.8905 | 2.2302 | 0.8995 | 2.9610 |
| Standard deviation | 0.0029 | 0.3886 | 0.0027 | 0.5071 | 0.0029 | 0.5038 | 0.0029 | 0.4687 |
| Minimum | 0.8646 | 0.7353 | 0.8753 | 1.0050 | 0.8814 | 0.6714 | 0.8939 | 1.8360 |
| 1 quartile | 0.8717 | 1.5352 | 0.8811 | 1.9410 | 0.8892 | 1.9131 | 0.9871 | 2.7050 |
| Median | 0.8732 | 1.8525 | 0.8825 | 2.1640 | 0.8910 | 2.2310 | 0.8992 | 2.9790 |
| 3 quartile | 0.8755 | 2.1422 | 0.8835 | 2.4900 | 0.8924 | 2.5742 | 0.9019 | 3.2300 |
| Maximum | 0.8789 | 2.3575 | 0.8880 | 3.1470 | 0.8949 | 3.0080 | 0.9050 | 4.0200 |

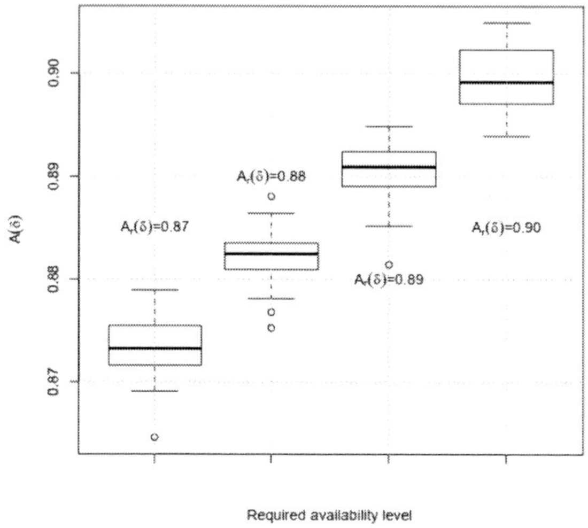


**Fig. 5.** Box-plot diagram of the technical objects availability value distribution.

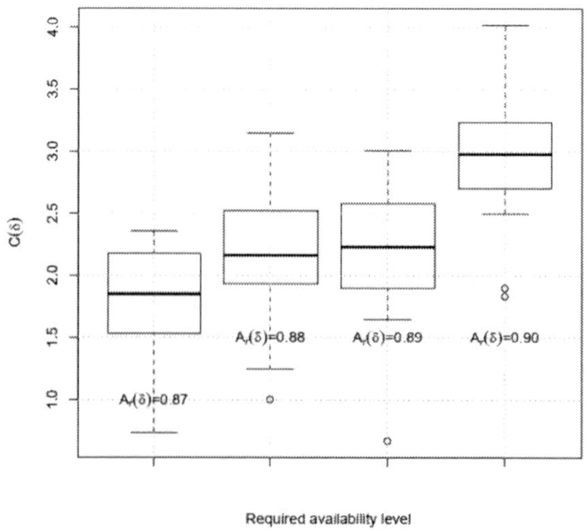


**Fig. 6.** Box-plot diagram of the unit income value's distribution.

## 5 Summary

Based on the obtained results it may be noticed, that in the analysed operating system, the increase of the means of transport's availability results in the increase of the unit income's value. It is caused mainly by the fact, that the tested vehicles belonged to the group of the newest vehicles (age of the vehicles from 2 to 3 years), and thereby these were the vehicles in case of which there was no necessity to incur considerable costs for their treatment.

Analysing the obtained results it may be noticed, that within the scope of the considered value of the availability required $A_r(\delta)$ from 0.87 to 0.90, together with the increase of availability, there occurs the increase of the mean unit income's value $C(\delta)$ respectively from the value 1.8034 to the value 2.9610 [PLN/h]. In the presented example it may be noted, that the availability increase from the level from 0.87 to 0.89, results in a slight increase of the unit income's value respectively from 1.8034 to 2.2302 [PLN/h], that is on average the increase of the availability factor's value by 0.1, results in the increase of the unit income by the value of 0.2134 [PLN/h].

Obtaining of the highest, analysed level of technical objects availability (for $A_r(\delta) = 0.90$), results in the highest change of the unit income's value $\Delta C(\delta) = 2.9610 - 2.2302 = 0.7308$ [PLN/h]. Generally, low level of the unit income in the studies system of means of transport's operating results from the fact, that the studied system is the strategical enterprise owned the city, and its task it is not to generate profit but only ensuring an effective realization of the transportations tasks at a given area.

The simulation model presented in the article, is a partial result of the conducted studies, the purpose of which is to develop a complex method of the means of transport operating process's steering with the use of decisive semi-Markov processes and nondeterministic methods of optimum solutions' assigning (evolutionary algorithms).

## References

1. F. Grabski, *Semi-Markov processes: Applications in system reliability and maintenance*. (Amsterdam, Elsevier, 2014)
2. F. Grabski, *Analiza ryzyka w decyzyjnych semi-markowskich modelach procesu eksploatacji. Risk analysis in decision-making semi-Markov models of operation process,* (XXXVIII Zimowa Szkoła Niezawodności. 38th Winter Reliability School, 2010, Szczyrk)
3. L. Knopik, K. Migawa, A. Wdzięczny, *Profit optimalization in operation systems*. (Polish Maritime Research, 2016)
4. V. G. Kulkarni, *Modeling and analysis of stochastic systems* (New York, Chapman & Hall, 1995)
5. K. W. Lee, *Stochastic models for random-request availability* (IEEE Trans, Reliability, 2000)
6. P. Marbach, T. N. Tsitsiklis, *Simulation based optimization of Markov reward process* (IEEE Trans. Automat. Contr, 2001)
7. K. Migawa, A. Neubauer, Ł. Muślewski, B. Landowski, *Method of availability control in the transport system using simulation model of operation and maintenance process* (Journal of KONBiN, 2017)
8. K. Migawa, L. Knopik, A. Neubauer, D. Perczyński, *Simulation of the model of technical object availability control* (Engineering Mechanics, 2017, Czech Republic)
9. K. Migawa, L. Knopik, S. Wawrzyniak, *Application of genetic algorithm to control the availability of technical systems* (Engineering Mechanics, 2016, Czech Republic)
10. Ł. Muślewski, K. Migawa, L. Knopik, *Control of technical objects operation quality with the use of simulation modeling PE* (Risk, Reliability and Safety: Innovating Theory and Practice: Proceedings of ESREL, 2016)

11. M. Szubartowski, K. Migawa, L. Knopik, M. Gniot, *Simulation model of risk evaluation in transport system* (Proceedings of 58th ICMD, 2017, Prague, Czech Republic)
12. M. Woropay, K. Migawa, *Markov model of the operational use process in an autonomous system.* (Polish J. Environmental Studies, 2007)
13. R. Kostek, P. Aleksandrowicz, *Simulation of car collision with an impact block* (11th International Congress of Automotive and Transport Engineering, Pitesti, Romania, 2017)
14. R. Kostek, P. Aleksandrowicz, *Simulation of the right-angle car collision based on identified parameters* (11th International Congress of Automotive and Transport Engineering, Pitesti, Romania, 2017)

MATEC Web of Conferences **302**, 01017 (2019) https://doi.org/10.1051/matecconf/201930201017
*18th International Conference Diagnostics of Machines and Vehicles*

# Exploring the effectiveness of using devices that follow the Sun

*Janusz* Musiał[1,*] , *Joanna* Wilczarska[1], *Oleg* Polishchuk[2] and *Andrii* Ramskyi[2]

[1]Faculty of Mechanical Engineering UTP in Bydgoszcz, Department of Vehicle Engineering, Al. Prof. S. Kaliskiego 7, 85-796 Bydgoszcz, Poland
[2]Khmelnitsky National University, Instytutska Street, 11, Khmelnytskyi, 29000, Ukraine

**Abstract.** The article is devoted to the study of the effectiveness of devices that follow the Sun. For this purpose, the construction of a two-axes tracker with an automatic system of steering the solar panels on azimuth during the light period of the day and with the possibility of manually adjusting the angle of rotation of the solar panels on the zenith, depending on the season is designed. For the proposed construction, the calculation of the angles of the zenith is made and the optimal angles are selected for manual adjustment of the seasonal positions of the system. The photo modules, which are installed stationary on the developed tracker system, are selected. Structural and electrical principle diagram of the stand was developed for determining the efficiency of solar power station, equipment was selected and its manufacture made. A solar power station monitoring system has been developed, which enables to automatically visualize parameters, perform data processing in real time, create databases and perform a comparative analysis of their operational efficiency in an automatic mode. Experimental researches have been carried out on determination of the generated electric energy by static and dynamic solar power stations, on the basis of which the efficiency of tracker systems of rotation of solar panels is confirmed. The use of dynamic systems compared to static allows to increase the production of electricity by 33% under the same conditions. But with poor illumination of the panels (rain, fog, cloudy weather), the tracker's effectiveness decreases and it does not exaggerate 5% when fully covered with the sky. Therefore, the use of tracking systems in the absence of direct sunlight on the surface of photo modules is not feasible.

## 1 Introduction

Solar energy is one of the promising directions of renewable energy sources that have been actively implemented in recent years in the world industry and life [1, 2]. Solar energy systems and devices allow efficient production of electric and thermal energy, thanks to the natural, practically inexhaustible possibilities of light radiation of the Sun with the least impact on the environmental state [3].

There is a colossal amount of solar energy that comes on the surface of the planet Earth. At its power, the weekly solar radiation exceeds all the world's known reserves of uranium, oil and coal. One of the benefits of solar power is that it is environmentally friendly. In its production, carbon dioxide is not formed, as, for example, at thermal stations. Unlike atomic power plants, it is completely radiation-safe. When it is made, waste is not formed that requires further utilization (slag, radioactive waste). That is why at present solar power is one of the most actively developing. For example, the total power of all solar power stations in the world in 2010 was 40.3 GW, and in 2022, according to Solar Power Europe,

* Corresponding author: Lukasz.Muslewski@utp.edu.pl

it will be 621.7 GW, which is more than twice the power of all existing nuclear power plants.

Improving the technology of producing electricity from the energy of the sun leads to the fact that over time, solar energy will become the cheapest among other alternative sources of energy. That is why studies conducted in the field of solar energy are relevant.

## 2 Problem statement

The difficulty of using the solar energy with a photoelectric panel is created by seasonal and daily unevenness of solar energy. This unevenness of energy flow can be caused by the Sun, weather conditions, as well as the daily course of the Sun in the sky. Effective use of solar energy by a photoelectric panel depends on its orientation in space. This will ensure a minimum deviation of the angle of incidence of sunlight to the normal photovoltaic panel during the year and the light period of the day [4].

There are two main types of solar panel mounting systems: static and dynamic. The static system is characterized by the fact that the orientation of the modules and the angle of inclination with respect to the Sun are unchanged. It is logical that solar modules should be maximally illuminated during daylight hours, so they should be oriented toward the south. If for some reason there is no such opportunity, then the solar panels can be installed on the south-west or south-easterly direction. The performance of the solar panels is also influenced by the angle of inclination, which is set depending on the location of the location of the solar power station [3].

One of the most effective methods for increasing the electrical performance of photo modules is the use of dynamic mounting systems and the use of devices that track the Sun (trackers). The advantage of moving trackers is that the solar panels placed on them automatically move in the sun during the day and change the angle of inclination depending on the season, which can significantly increase the efficiency of the station and increase the production of electricity compared with fixedly fixed solar panels [3, 5].

The choice of a tracker type depends on many factors, including the size of the installation, electrical parameters, land restrictions, latitude and local weather conditions.

Among the most commonly used panel orientation systems, the following can be distinguished: continuously-tracking with the position sensors of the Sun and continuous or discrete-tracking with program control [4]. In addition, they can be made by one- or two-axis position at the axis of rotation [3].

Trackers with one axis of rotation – single-track trackers have one degree of freedom that acts as an axis of rotation. The axis of rotation of a single-axle tracker is usually oriented along the north-south axis. It is possible to orient them in any coordinate directions using tracking algorithms.

Trackers with two axes of rotation have two degrees of freedom, which act as axes of rotation. These axes, as a rule, are not connected with each other, but work together. The axis that is fixed in relation to the Earth can be considered as the main axis. Another axis can be considered as secondary. Such trackers usually have modules oriented parallel to the secondary axis of rotation. They allow taking the optimal amount of solar energy due to their ability to monitor the sun in both vertical and horizontal directions.

Summing up the advantages of dynamic systems for fixing photovoltaic modules, it is necessary to develop a tracking tracker system that would significantly improve the productivity of solar panels and explore its parameters.

## 3 Analysis of recent research and publications

In connection with the development of electric drives and power electronics, photomultipliers increasingly use dynamic automatic systems for directing photo modules on the Sun, which provide an increase in their productivity, from different sources, from 25 to 47% [6, 7]. The advantage of such guidance systems is that the solar cells placed on them automatically move around the Sun throughout the day and change the angle of inclination depending on the season. This significantly increases the power generation compared to stationary (immobile) fixed solar panels due to the fall of sunlight at an angle of 90° to the plane of the panel during the light day, leading to the generation of maximum energy, which in turn provides the maximum energy efficiency of solar panels.

Analysis of technical literature has shown that the greatest energy generation occurs when the tracker system turns the solar panels to a maximum angle in the azimuth plane. The zenith angle of the slope of photo modules, depending on the time of year, does not significantly affect the production of electric energy. Therefore, it can be set the same for all seasons as optimum for a certain area, or regulate manually. Using an additional actuator to move the solar panel azimuthally leads to an additional cost of electricity for its power supply.

## 4 Presenting main material

The design of the tracker system of solar panels (Fig. 1) is proposed, which consists of: a carrier design of the tracker; azimuth orientation system; actuator; photoelectric sensor and microcontroller to control the orientation process [3].

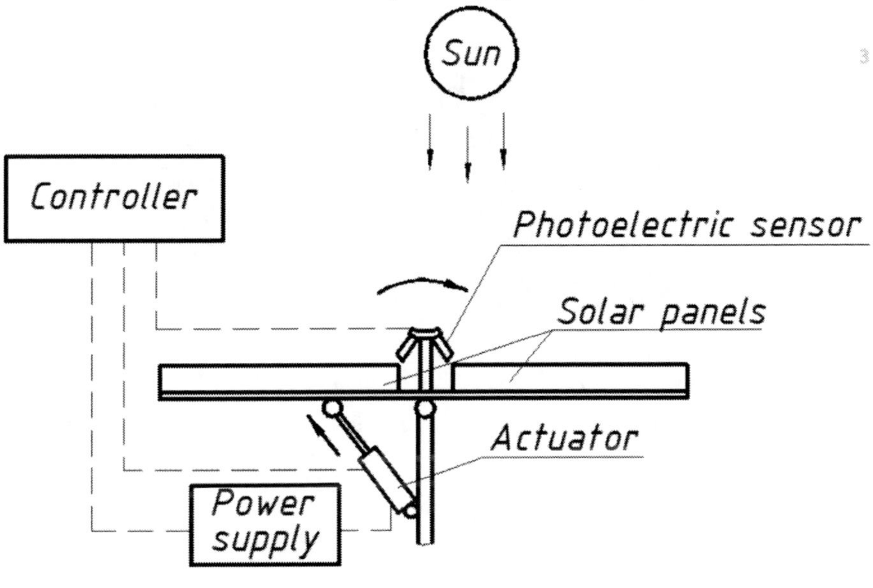


**Fig. 1.** Design scheme of the tracker system of solar panels.

A kinematic scheme of the tracker mechanism was developed (Fig. 2, 3) and the sizes of all the components included in it were calculated. A double-axes tracker for rotation of solar panels was designed on the basis of the obtained data.

The technical capabilities of the design of the tracker allow rotate the panels to the maximum angle in the azimuth plane. The zenith angle is manually adjusted by means of traction. To do this, a calibration is applied on the frame, which allows the seasonal positions to be displayed by the labels. This made it possible to significantly reduce the cost and simplify the design, which in turn reduced the payback period of the plant without significant losses in the efficiency factor.

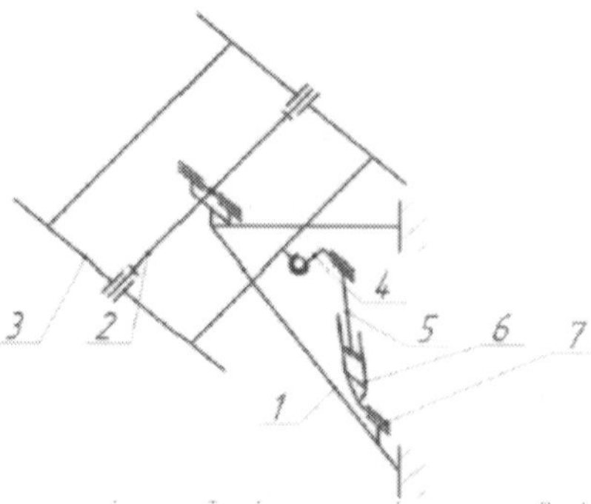


**Fig. 2.** Kinematic scheme of the tracker:
1 - a frame with an undercut; 2 - axis of the tracker; 4 - turntable; 5 - drive socket; 6 - drive housing; 7 - the hinge.

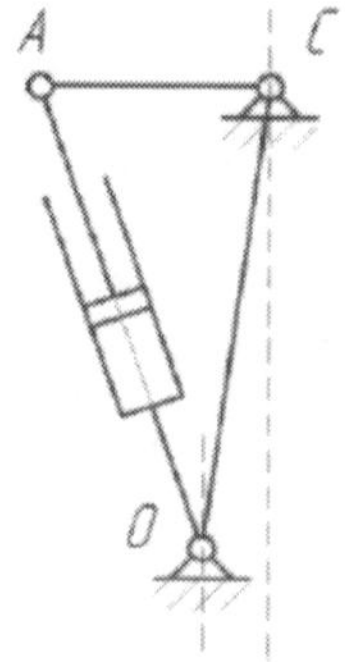


**Fig. 3.** Kinematic scheme of the executive mechanism.

For the purpose of maximum illumination in the light period of the day static and dynamic systems were installed strictly to the south with the same angle of inclination in the plane of the zenith. The general view of the installed photo modules is shown in Fig. 4. In the static and dynamic solar power stations, two 150 W solar panels are used each.

**Fig. 4.** General view of the installed photomodules: 1 - static; 2 – dynamically.

Output data for determining the optimal angles in the zenith plane is determined using a specially designed program. To do this, using the GPS module, the coordinates of the tracker's location were determined: the geographical latitude of the defined point is H 49.4066; geographical longitude is E 26.9639.

The actuator, illumination sensor and microcontroller for controlling the process of orientation of the modules were also selected.

To study the efficiency of solar power stations, a structural diagram of the stand, presented in Fig. 5, was developed. The block diagram in the center has a module for collecting and processing information Arduino Uno [8]. After processing, the signal is received through the Wi-Fi module, the Internet network on the computer. Signals are fixed by current sensors. Signals from sensors and voltage dividers enter the analog inputs of the data processing module, digitized, processed and transmitted to the computer in the way described below.

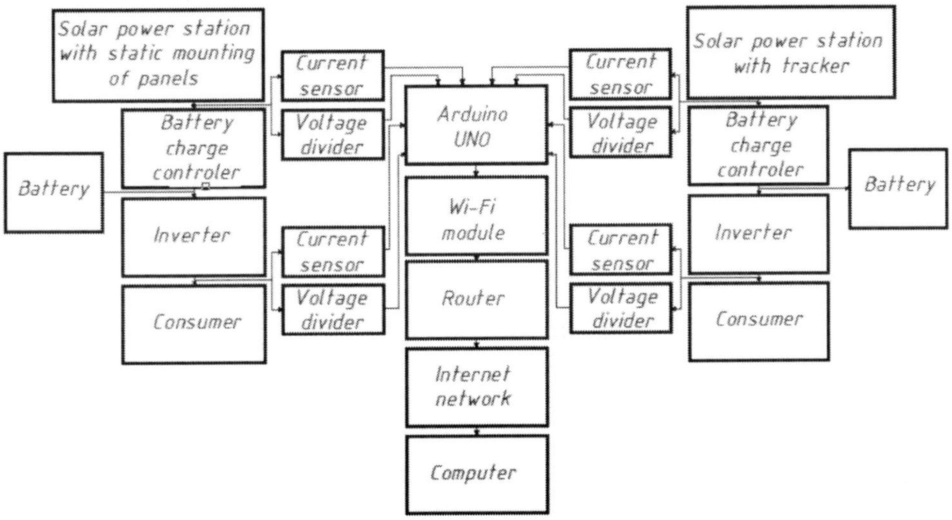


**Fig. 5.** The block diagram of the stand for the study of the efficiency of solar power plants.

On the basis of the structural scheme, an electric circuit diagram of the stand was developed and the necessary equipment was selected. The general view of the experimental stand is shown in Fig. 6.

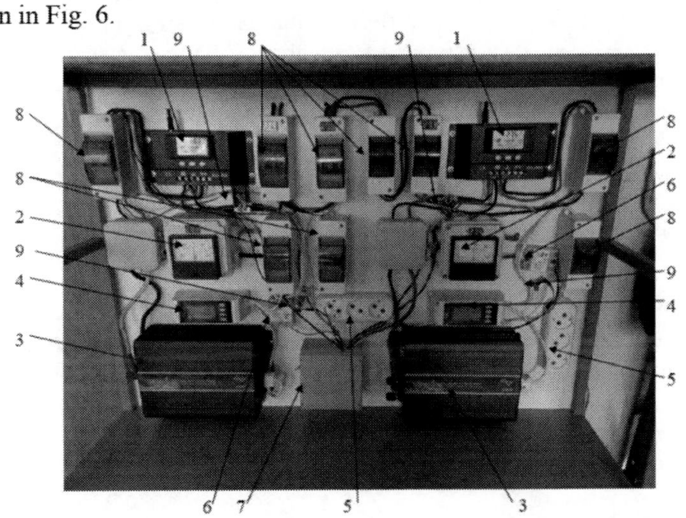

**Fig. 6.** General view of the experimental stand: 1 - battery charge controller; 2 - ammeter; 3 - inverter; 4 - Controller (contains a built-in display for displaying current and voltage output); 5 - socket for connecting the load; 6 - solid state relay; 7 - module for collecting and processing information with the built-in power supply; 8 - electric switch with overload protection and short circuit; 9 - current sensor.

A monitoring system for two independent solar power stations was also created. The created system of research of the effectiveness of solar panels with tracker and stationary panel placement system enables to automatically visualize parameters, perform data processing in real time, form databases and perform a comparative analysis of their operational efficiency. It also allows to remotely track settings that are fixed from anywhere in the world [9-11]. The developed system makes it possible to compare the efficiency of solar panels. The WEB page of the developed monitoring system is shown in Fig. 7.

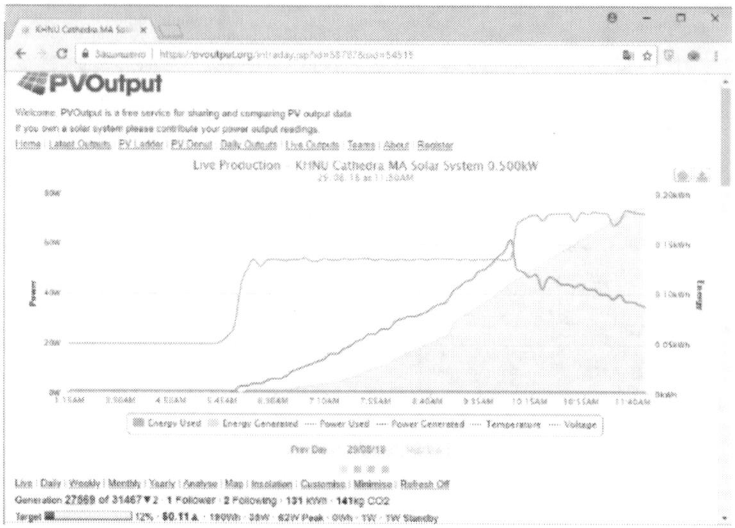


**Fig. 7.** WEB page of the developed monitoring system.

# 5 Results of experimental studies

Experimental studies were carried out in sunny and cloudy weather. Two days of sunshine, one sunny day and several days with different illumination of light day were chosen for the experiment. To ensure equal conditions of measurement, the solar panels of the two power stations were nearby, but the shade from the panels did not fall on one another. In the course of experimental research, the tracker system and data collection system worked in automatic mode. Data from photo modules was automatically removed and transferred to the server for plotting.

For the estimation of efficiency in numerical form, formula (1) used the percentage difference between two numbers:

$$a > b = ((a - b)/a) \cdot 100\% \quad (1)$$

where: $a$ is the value of the first number; $b$ is the value of the second number.

As a result of calculations, we obtain the percentage value that shows how one of the solar power stations is more efficient than the other over a certain period of time.

The first experiments were conducted on sunny and almost cloudless days (26 and 27 June 2018). Measurement of values was carried out at intervals of 5 minutes.

Fig. 8, 9 shows the received graphs from the WEB-page for two power stations.

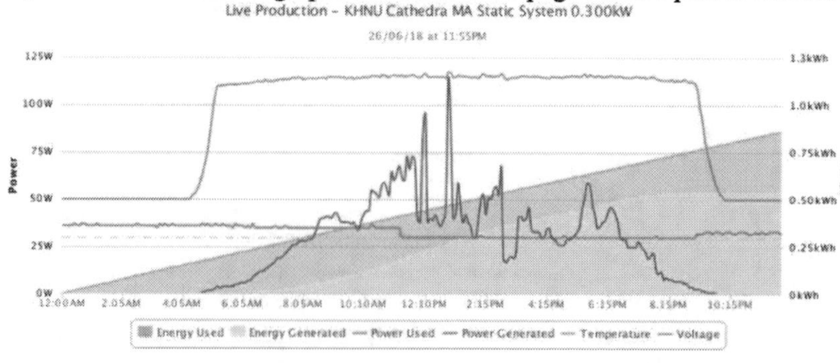


**Fig. 8.** Graphs of produced and consumed electric energy of a static solar power station as of 06.26.2018.

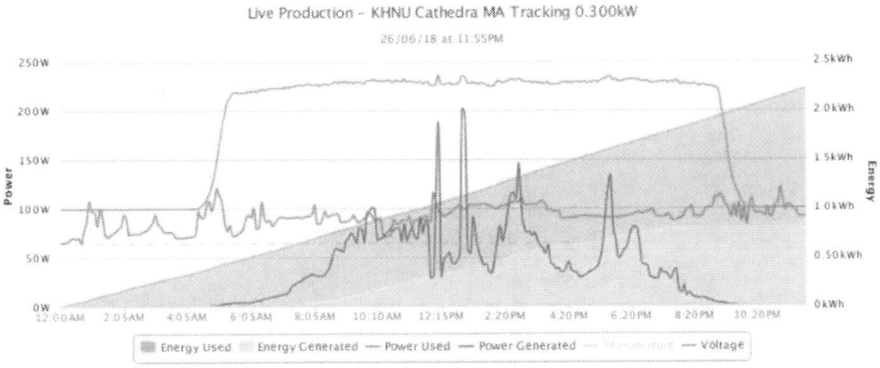


**Fig.9.** Graphs of produced and consumed electric energy of tracker solar power station as of 06.26.2018.

Total light day static system produced 0.537 kWh. The peak power at 12:55 am was 112 Watts. Over the course of the day a tracking system produced 0.806 kWh. The peak power at 12:55 am was 202 W, respectively.

For comparison, the combined graphs of electricity produced by two power stations were plotted (Fig. 10).

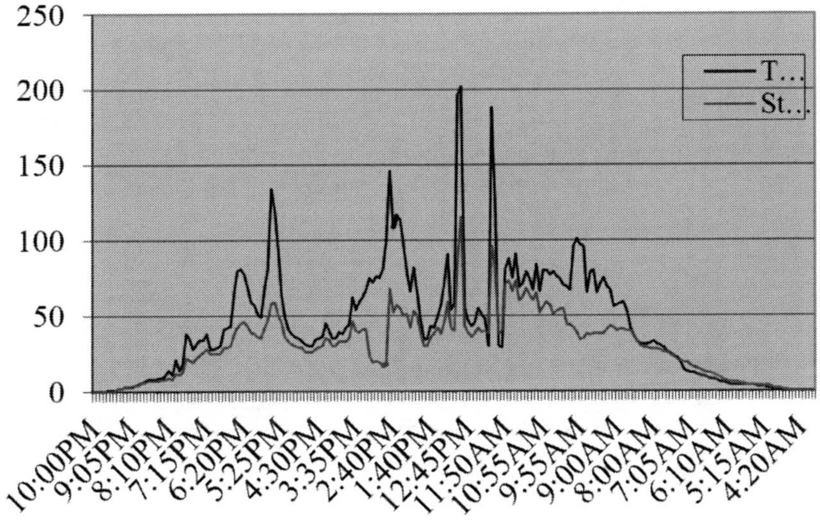


**Fig. 10.** Coherent graphs of produced electric energy of tracker and static solar power stations on 06.26.2018

Using formula (1), the effectiveness of the tracker system was calculated, which was 33.37%.

Next, experimental research was carried out on the following day (June 27, 2018) with a lower ambient temperature and solar activity, and the second half of the day was generally cloudy.

Having analyzed the received graphs, it was found that the static power station produced 0.336 kWh, and the tracker one 0.412 kWh. Maximum peak power reached 80 Watts and 167 Watts, respectively. Maximum solar activity was around 12 o'clock in the day.

The combined graphs of the produced electric power of both power plants are shown in fig. 11.

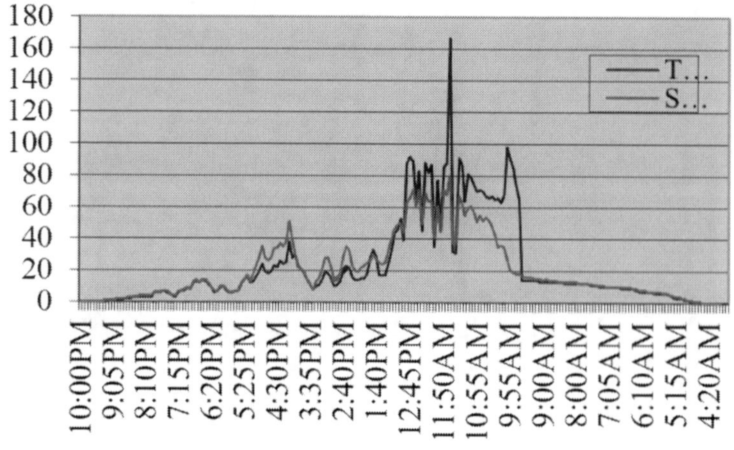


**Fig. 11.** Combined graphs of electric energy produced by tracker and static solar power plants on June 27, 2018.

The calculations carried out by the formula (1) show that on June 27, 2018, the tracker solar power station was 11.4% more efficient than the static.

The next experiment was conducted on September 13, 2018. On that day it was gloomy, there was no direct sunlight on the panel. The ambient temperature ranged from 16 to 20° C.

The obtained results (Fig. 12) showed that on that day, tracker and static solar power stations produced almost the same amount of electrical energy – 424 and 403 kWh, respectively. This is due to weather conditions, because the clouds in the sky prevented direct sunlight, and the illumination of the panel almost did not depend on the turn of the photo module.

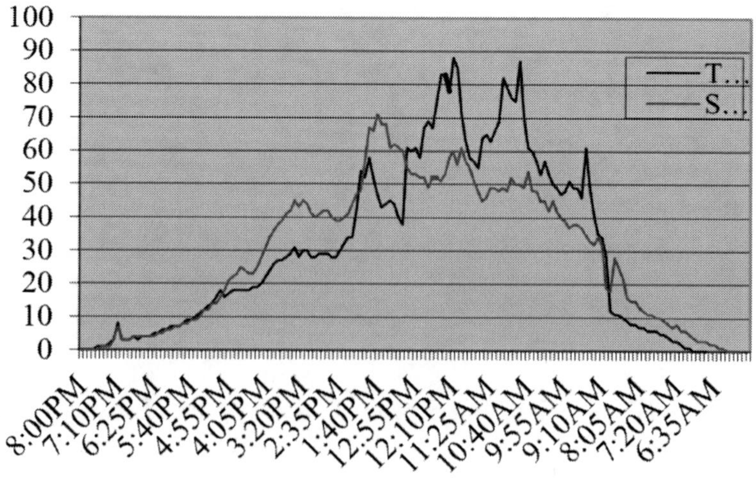


**Fig. 12.** Combined graphs of electric energy produced by tracker and static solar power station as of 09.13.2018.

The effectiveness of the tracker system on a cloudy day was 4.95%. The following experiment was conducted within three days from September 14, on September 16. These days were sunny, sometimes it was cloudy, but sunny activity was quite high. This experiment allowed us to evaluate the effectiveness of the tracker system over a long period of time. The largest amount of electricity was generated by solar power plants for these three days, 14.09.2018. Examples of the graphs received for this day are shown in Fig. 13, 14.

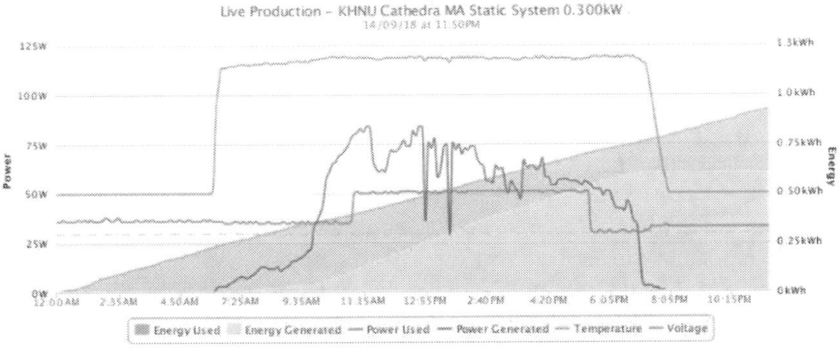


**Fig. 13.** Graph of produced and consumed electric power of static solar power station as of 09.14.2018, capacity 0,603 kW/h.

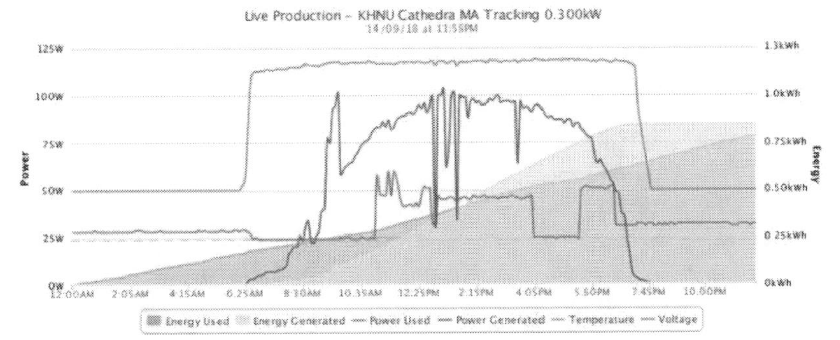


**Fig. 14.** Graph of produced and consumed electric energy of tracker solar power station as of 09.14.2018, capacity 0,841kW/h.

According to formula (1), the efficiency of the tracker power stations were compared with the static for 3 days. As a result of the calculations, it was 29.1%.

## 6 Conclusions

As a result of this work the design of a two-axis tracker with an automatic system of directing solar panels on azimuth during the light period of the day and with the possibility of manually adjusting the angle of rotation of solar panels on the zenith, depending on the season is developed. For the proposed design, the calculation of the angles of the zenith is made and the optimal angles are selected for manual adjustment of the seasonal positions of the system.

The developed solar power monitoring system is convenient at operation, enables to automatically visualize parameters, perform data processing in real time, form databases and perform a comparative analysis of their operational efficiency.

As a result of the experimental research on the determination of the produced electric energy of static and dynamic solar power stations, the efficiency of the tracking systems for the rotation of the solar panels was confirmed. The most effective tracker systems are in sunny weather. They can produce 33% more electrical energy than static under the same conditions. But with poor illumination of the panels (rain, fog, cloudy weather), the tracker's effectiveness decreases and it does not exaggerate 5% when fully covered with the sky. Therefore, the use of tracking systems in the absence of direct sunlight on the surface of photo modules is not feasible. In this case, the electric energy consumed by the tracker's drive may be greater than that which is additionally produced when solar panel follows the Sun.

## References

1. N.M. Mkhitaryan, S.O. Kudrya, V.F. Reztsov. Potential and prospects for the use of renewable energy in Ukraine Alternative energy and ecology, № 8. P.150-163, (2011)
2. A.O. Kasych, Y.O. Lytvynenko. Factors for the development of alternative energy in modern conditions. Economics and Society, №12. P.93-99, (2017)
3. Danko V.M., S.V. Smutko., Polishchuk O.S. Development of the design of the tracker system for solar panels. Herald of Khmelnytskyi National University, №1. P.232-235, (2017)
4. Boyarchuk V., Syrotyuk V., Galchak V., Syrotyuk S., Boltyansky B. Development of an experimental stand for the study of the effectiveness of the application of devices that follow the sun. Herald of Lviv National University: agroengineering research, №17, P.286-293, (2013)
5. V.P. Kozhemyako, O.G. Dombrovsky, V.F. Zherdetsky, V.I. Malinovsky, G.V. Prytulyak. An analytical review of modern technologies of photovoltaic transformers for solar power. Optoelectronic devices and components in laser and energy technologies, P.141-157, (2011)
6. Gevorkian P. Alternative energy systems in building design / P. Gevorkian. – NewYork: McGraw-Hill, – 522 p., (2010)
7. Gavin D.J. Harper. Solar Energy Projects for the Evil Genius / Gavin D. J. Harper – New York: McGraw Hill,. – 196 p., (2007)
8. W. Sommer. Arduino/Freeduino microcontroller programming. - SPb.: BHV-Petersburg, 256 p., (2012)
9. T. Kałaczyński, M. Łukasiewicz, J. Musiał, R. Polasik, M. Szczutkowski, N. Dluhunovych, J. Wilczarska, T. Kasprowicz, Analysis of the diagnostic potential research thermovision in the technical state of combustion engine injectors assessment, Engineering Mechanics, 1805-8248, 357-360, (2018)
10. A. Troszyński, M. Liss, Possibilities of the electrochemical drilling process's (ECM) vibroacoustic analysis, Matec Woc, 182, 1-8, (2018)
11. M. Styp-Rekowski, E. Mańka, M. Matuszewski, M. Madej, D. Ozimina, Tribological problems in shaft hoist ropes wear process. Industrial Lubrication and Tribology, **67**, (2015)

# Analysis of the impact of an operator's workstation on the safety of functioning of selected road transport systems

*Łukasz* Muślewski[1],*

[1]University of Science and Technology in Bydgoszcz, Faculty of Mechanical Engineering, Al. prof. S. Kaliskiego 7, 85-796 Bydgoszcz, Poland

**Abstract.** Modern times are characterized by fast economic and social development. However, not all spheres of life have developed in a sustainable manner, for example some elements of anthropotechnical systems are still far from being sustainable. This results in low safety level of these systems involving high risk of injury or death caused by road accidents. Transport systems of the type Human-Machine-Environment <H–M–E>, where undesired events are caused by human behavior, belong to such systems. The author of this study has addressed the issues connected with the impact of man-made objects on their operation safety. The objects in question are selected transport systems and the study subject is assessment of the equipment operator's behavior. The main goal of this study is to provide assessment and analysis of the impact of transport means ergonomics on the quality of transport services they provide. The research object are delivery trucks with the maximum vehicle weight up to 3.5t. The first part of the study contains identification of the factors and threats which affect functioning of transport systems. Further, ergonomics criterion is defined and the factors affecting ergonomics of selected transport means are provided. A questionnaire survey was performed with participation of professional drivers who were asked to identify the factors which have an adverse effect on the safety of the transport process as well as factors that increase the drivers' comfort of work. Basing on this, an analysis of the results was performed and negative aspects involved in driving ergonomics were identified.

## 1 Introduction

The problems connected with the safety of transport systems is an interdisciplinary issue and applies to transport psychology, theory of systems, ergonomics, transport, maintenance and others [1,2].

Socio-technical systems ⟨H–T–E⟩, including transport ones, are characterized by similar features regardless of the industry they are used in. The level of their safety is affected by the operator's behavior (human factor), the technical object (transport means) and the impact of the environment. It needs to be highlighted, though, that it is the human factor that is considered to have the highest influence on a transport system functioning safety[3,4].

The concept of factor, that is, a quantity being its theoretical image, is one of the notions of the science idealizing concepts [5]. Hence, a factor can be interpreted in the mathematical form defining its relations. Human factor as an element affecting a system is a special case of a factor. According to work [6], the concept of human factor is a design and engineering of the human-machine-environment systems to improve people's

* Corresponding author: Lukasz.Muslewski@utp.edu.pl

behaviors. The concept of 'human factors' is often referred to as a scientific discipline like ergonomics. Study [7] provides an analysis of the human factor concept, thus for the needs of this study, a definition was accepted which assumes that a human factor is a general term commonly used as a name of a specialized discipline which deals with the relations between a man an a machine. It focuses on the problems of perception, psycho-physics, decision making and other aspects involved in the information processing. Sometimes it applies to such elements as: equipment, physical environment, tasks and persons who perform the work [8-10]. A system 'human – machine" is a relatively closed system, i.e. the processes involved in the system operation are of cyclic character and always return to the initial point which does not rule out entrances into it and exits from it including cooperation with the environment [11]. The human factor is often identified with a human, however, such simplification does not provide a full picture of the human impact as it should always be considered in a situational contexts which affect human reactions and predispositions such as: temperament, aggression, feeling of control etc. [7].

The design process involves integrating technical structures with humans into one biotechnical system 'human – to – environment". Therefore, all machines need to be equipped with specially constructed workstations for the operators in order to provide the system with safe operation. If it comes to transport means operation systems, this is the machine operator who is usually responsible for collisions and accidents [12]. A respective directive includes major requirements regarding safety and health protection in relation to all machines [13].

## 2 Identification of the factors affecting safety of transport system operation

Providing technical systems with high safety level is a problem tha occurs in many branches of the national economy. Road traffic transport system are socio-technical systems which need to be provided with ensuring high safety level because their operation is largely affected by people who are operators of technical objects, passengers of transport means and road traffic participants (bikers, side walkers, others) [14].

The concept of ***transport safety*** *is understood as the system ability to operate in a given environments with no accidents or undesired road events.* In work [15] the authors conclude that the science regarding safety, also referred to as safety engineering, addresses the methods for loss prevention which can be used for verification of the system safety control decisions, basing on the results of socio-technical models tests. An operation system can be considered rational only when it contributes to, optimal in terms of accepted criteria, most often economic, achievement of set goals. [16]

The below listed environmental factors have an influence on a driver's behavior in the road environment[17]:

- other vehicles that make driving more difficult,
- pedestrians on the pavements who distract attention,
- pedestrians entering the road,
- bikers on the road,
- dangerous behavior of other drivers,
- animals near the road,
- time of waiting in line to enter an intersection,
- standing in a traffic jam,
- frequent stops in front of traffic lights,
- bumps in the road,
- water, mud, snow,
- slippery surface,
- weather, climate,

- ads in the road proximity, shop displays,
- insufficient road lighting,
- blinding by the sunlight or other vehicles,
- invisible horizontal signs,
- poor visibility of the road, vertical signs and traffic lights,
- confusing road traffic rules,
- to many signs and too much information,
- unclear marked intersection.

According to the author [17], the features of the road environment have a significant impact on the safety margin to be accepted by the drivers, hence, on the number of road traffic victims.

According to the authors of work [18], the factors affecting safe driving can be divided according to four criteria due to:

- external environment of the vehicle:
  - quality of roads and their marking,
  - atmospheric conditions,
  - road traffic volume,
  - time – day/night,
- workstation:
  - workstation ergonomics,
  - physical factors, i.e. noise, vibrations, temperature,
  - technical state of the vehicle,
- the driver:
  - psycho-physical fitness,
  - personality traits,
  - age and experience,
  - fatigue and sleep deficit,
  - health condition,
- work organization:
  - length of distance to cover,
  - time of travel,
  - stops to relax.

All these factors have an influence on the driver's behavior which affects safety of the passengers and cargo and can pose a threat to lives of the passengers, other road users and their own. The driver's behavior and actions can be divided into three categories [19]:

- logic,
- illogic,
- irrational.

The behavior of drivers is affected by emotional processes, character traits, temperament, motivation, attitudes which is closely related to the driver's age and his/her job experience. Psychological aspects of the driver's personality include [19]:

- thinking,
- memory,
- perception,
- attention.

Physiological predispositions of drivers including vision, hearing, responsiveness, reaction time, alertness, drowsiness and fatigue which depend on the age and gender play a very important role in driving as well [15, 17, 19]. The authors of work [4] have analyzed the impact of ergonomic risk and muscle-spine related ailments which accompany a driver of a passenger car while driving and have indicated which health problems may occur.

In its further part, the study deals with the problems of ergonomics of different types of vehicles which is a significant criterion for assessment of transport systems functioning safety.

## 3 Analysis of the problems connected with drivers' workstation ergonomics

The name ergonomics comes from Greek. *Ergon* work and *nomos* natural laws. Ergonomics is a science which combines human psycho-physical abilities with operation of machines and devices [20].

Ergonomics is referred to as an applied science. It focuses on human needs when designing transport means to make them adjusted to the operator and the passengers [21].

The beginnings of the concept of ergonomics date back to the period of creation of the first tools. Homo habilis that is, a skillful man, was probably the first to pay attention to the aspect of ergonomics while creating new tools. After the II world war, ergonomics developed to provide safe working conditions. It aimed at optimizing maximum muscle loads and force values to be used by a worker to do his/her job [22].

Ergonomic aims at adjusting tools, machines and devices to the physical and psychical needs of the man. It focuses on the human needs rather than the economic effect [4, 23].

The major assumption of ergonomics is to provide solutions that are most effective in terms of economy and ecology and are involved in job efficiency, comfort and safety improvement [7].

Ergonomics refers to such scientific disciplines as: anthropology (deals with measurements of human body to be used in creation of a workstation), work physiology (assessment of phenomena that occur in work in order to prevent fatigue and tiredness), occupational safety and health (prevents factors that cause diseases),work psychology (addresses two aspects: adaptation of a human to work, adjustment of work to a human , and adaptation of a human to other humans) [24, 25, 26].

Analyzing the factors conditioning drivers' job it needs to be emphasized that they involve many aspects affecting their health which in turn has an impact on accomplishment of transport tasks. In their daily work the drivers have to face the internal and external factors which directly affect their behaviors in the road traffic environment [1].

## 4 Ergonomics of selected transport means

In view of the above, it needs to be highlighted that ergonomics of a driver's workstation is one of the factors that have a significant impact on the transport means operators' work quality which translates directly into safety of transport services to be provided by them.

### 4.1 Survey

The study includes the results of a survey on the subject of workstation economics for drivers of delivery trucks up to 3.5t.

The survey was carried out for a group of professional drivers, classified in terms of age, gender, driving license class and the age of their vehicles. Persons who were employed in transport companies as users of vehicles were statistical units of the survey. The surveyed group included 50 respondents. The survey subject was the vehicle ergonomics and the main goal of the survey was assessment of ergonomic conditions while driving the vehicle. The survey consisted of seven questions, both open and closed ones. Each questionnaire was marked from 1-50.

The analyzed group of respondents included 45 men and 5 women. The age of fifteen respondents was 41- 50, twenty of them were (31-40 ), 8 were more than 50 years old and seven from 21- 30. It needs to be noted that those younger than 21 were not included in the survey. In terms of job experience the respondents were divided into groups according to the period of employment. Most of the respondents had been employed for more than 5 years (35), 7 for 3-5 years, 6 for 1-3 years, whereas 2 shorter than for one year. The next aspect taken into account was the vehicle used by the respondents. They were divided into two types: up to do 3.5 t, with no trailer and with a trailer. The first group included 39 vehicles, whereas, the second 11 vehicles. The last division criterion of the respondents was the age of their vehicles. It consisted of five time periods. Most drivers (16) used vehicles manufactured in (2006- 2010), 19 used trucks manufactured after 2010, 9 between (2001-2005), 5 vehicles from (1995-2001), one driver used a vehicle manufactured before 1995. The survey questions are presented in table 1.

**Table 1.** The survey questions.

| No. | Question |
|---|---|
| 1. | What is your opinion about the comfort of driving according to grading scale from 1(poor) to 5 (excellent)? |
| 2. | Which element of the drover's cabin needs to be changed or modified and why (seat, steering wheel, gear change system, windshield mirrors, safety belts, pole, armrest, soundsystem others )? |
| 3. | Provide the assessment of difficulty in operation of the cabin equipment 1 (low) -5 high) and list the devices which you find most difficult to operate? |
| 4. | Which of the external factors and to what degree is the most annoying for you while driving the vehicle, use scale from 1 (the least ) to 5 (the most) (noise, precipitations, slippery road surface, limited visibility, traffic jams temperature, other road users)? |
| 5. | Which devices and to what degree according to the scale 1 (the least ) -5 (the most), make ypur work more enjoyable (fridge, television, roof-window, navigation, multi-functional steering wheel, air conditioning webasto, gas cooker, others)? |
| 6. | Which activity/manouver is difficult for you due tothe vehicle structure and/or equipment (parking, loading/unloading, compulsory standby, driving)? |
| 7. | Is it possible to improve ergonomics of your vehicle from the point of view of the operator? |

### 4.2 Survey results and their analysis

On the basis of the survey results carried out for a group of fifty delivery van drivers, the basic statistical indexes were determined for the obtained scores, which are presented in table 2.

**Table 2.** Mean values of the survey scores.

| | Questions | | | | | | |
|---|---|---|---|---|---|---|---|
| | 1 | 2 | 3 | 4 | 5 | 6 | 7 |
| Sum | 177.5 | 209 | 182.5 | 187.3 | 59.66 | 188.8 | 205 |
| Average | 3.55 | 4.18 | 3.65 | 3.75 | 1.19 | 3.78 | 4.10 |
| Median | 3.75 | 4.50 | 3.75 | 4.29 | 1.06 | 3.75 | 5.00 |
| Stand-dev | 0.99 | 0.55 | 1.40 | 0.96 | 0.89 | 0.31 | 1.94 |
| Coef-Var | 0.01 | 0.003 | 0.01 | 0.01 | 0.01 | 0.002 | 0.01 |
| Min | 1.25 | 2.5 | 1.25 | 2.43 | 0.00 | 2.5 | 0 |
| Max | 5 | 5 | 5 | 4.57 | 3.89 | 5 | 5 |

An analysis of the answers to the first question shows that it was given the two highest grades by the biggest number of the respondents. Hence, it can be said that a great majority of the drivers are satisfied with the comfort of driving their vehicles.

Number 2 was supposed to find out which element of the cabin equipment needs to be changed or modernized. Almost 50% of the respondents indicated the driver's seat. They claimed that: it is not comfortable, it often breaks down, its adjustment system is poor, is badly profiled, long driving causes pain in lumbar vertebra, there should be hydraulic adjustment, the foam inside the seat is of poor quality, a longer trip results in spine pain, head rests are not of sufficient quality, the seat should be better matched to the body. The second elements which was indicated by the drivers as not good enough were mirrors. They indicated that the mirrors: limit visibility, are not heated, generate the so called ' blind spot", they should be larger, their shoulders are too short, should be electrically controlled and they provide too small field of view. The next problem that was indicated by the respondents was the system od gear change and the type of gear lever 'A', which was mentioned by eight persons for each issue. Another element which was indicated by 7 respondents was the steering wheel. The drivers claimed that the steering wheel should be multi-functional and made of leather. As far as safety belts are concerned the respondents said that: they should be more elastic, they often get blocked and are too tight and 'pull'. Two drivers indicated the arm rest as an element that needs to be modernized because: it is too stiff, and there should be two armrests on both sides. The windshield has some characteristics which undoubtedly disturb the drivers in their work. These are: poor visibility, generating frost or steam as well as the fact that the wipers do not cover the whole surface of the windshield. The drivers had also objections concerning the sound system and loudspeakers and their configuration which in their opinion needed modification.

The third question consisted of two parts. The first part was about the level of difficulty in operation of the devices inside the driver's cabin and it turned out out that the drivers did not find it difficult to operate the cabin equipment, whereas 9 respondents had some difficulties in operation of the tachometer, use of navigation, reading the deck computer announcements, operation of the radio and temperature programmer.

A multiple choice question number 4 was connected with the most annoying external factors. Most respondents indicated the infrastructure and traffic jams (including congestions) as the most irritating factors. It is commonly known that traffic congestions occur in many towns particularly during rush hours. They are very disturbing due to the pressure of delivery time. Other annoying elements include: noise, precipitation and slippery road surface.

The fifth question was about the most useful devices fixed in the cabin such as: fridge, navigation, webasto, television, multi-functional steering wheel, gas cooker, window roof, air conditioning or others. According to the respondents the most useful devices are navigation and air conditioning. Television scored the least points. The respondents found navigation and air conditioning to be indispensable for driving as navigation significantly supports driving and air conditioning significantly increases the comfort of work.

The sixth question was about difficulties in activities involved in their job. Respondents chose a specific activity and gave reasons why it poses the most problems. Most respondents indicated loading and unloading to be the most disturbing activity, whereas, according to them, the least disturbing was driving itself. In their opinion loading and unloading take most time, there are not enough tapping points, the loading should be mechanical, some companies take drivers for persons responsible for loading and unloading the truck, besides they have to wait for too long and there are no side doors to enable unloading on the side or there is no special platform to be used for loading/unloading. Parking was chosen to be problematic as well because of: lack of parking sensors, lack of cameras, fear of damage or accident due to the vehicle large size and high length, or little parking space or limited visibility.

Question 7 was about the vehicle ergonomics improvement. The drivers could propose their own solutions to improve the work quality and efficiency. 39 respondents said that it was not possible to improve ergonomics in their vehicles, however, the other respondents proposed some solutions including: modification of the suspension, increasing the engine power, increasing visibility, montage of air conditioning, replacement of the standard steering wheel with a multifunctional one, automatic gear box, improvement of the driver's seat, reduction of noise and creation of more lockers.

Additionally, in order to check correlation between particular questions the correlation coefficient significance test was performed on the basis of the survey results. Since the population of the sample was n=50, hence it is statistically significant for the variables from the continuous distribution. Based on an analysis of the results it was found that only four pairs of variables have significant correlations in relation to each other, whereas their significance level is expressed by low values. It provides the basis to state that this set is not redundant.

## 5 Conclusions

Transport systems are specific types of socio-technical systems where more than 90% of the undesirable events are caused by fault of the human. The decisions of a vehicle operator, that is a driver, which are frequently affected by the ergonomic factors, have a significant impact on the vehicle. Many of these factors influence the driver's behavior which in case of fast reaction can lead to a collision or an accident. To provide appropriate assessment of the operator's behavior in a situation of a higher risk it is necessary to investigate the impact of the vehicle ergonomic conditions.

This study provides an analysis of an operator's – a supply truck driver's workstation ergonomics. A survey was used to identify factors that have an adverse effect on the quality of the transport process. The impact of selected internal and external factors on drivers' work ergonomics was assessed. The problems and factors which could contribute to ergonomic improvement in the transport means they use have been analyzed. A statistical analysis and the correlation coefficient significance test was made as well. As the level of the correlation significance value was particular questions are disjoint and do not reflect the information which is included in the other questions. There is a need to conduct further research on this subject and compare the results. The group of passenger car or supply trucks with more than 3.5t.capacity drivers or drivers of other transport means should be surveyed as well to better match the machines and devices with their operators' physical and psychological traits.

## References

1. Ł. Muślewski, B. Landowski, N. Mackiewicz, M. Pająk, Research and analysis of work ergonomics of selected transport means operators. International Automotive Conference, Materials Science and Engineering, 1757-8981, (2018)
2. M. Pająk, Ł. Muślewski, B. Landowski, A. Grządziela, Fuzzy identification of the reliability state of the mine detecting ship propulsion system. Polish Maritime Research 1(101) Vol.**26**, 55-64 (2019)
3. D. Koradecka (red), *Occupational safety and health*, (CIOP – PIB, Warszawa 2008)
4. R. Makarowski, *Risk and stress in sports aviation,* (Difin Warszawa 2010)
5. Transport Research Laboratory, Construction logistics and cyclist safety, Technical report (2012)
6. P.G. Dempsey, M.S. Wogalter, P.A. Hancock, What's in a name? Using terms from definitions to examine the fundamental foundation of human factors and ergonomics science (2000)
7. Ł. Muślewski, Evaluation method of transport systems operation quality. Polish Journal of Environmental Studies. Vol. **18**, No. 2A (2009)
8. D. Starkowski, K. Bieńczak, W. Zwierzycki, Carv transport in Poland and abroad. Compendium of practical knowledge. Vol III. Driver's work environment.

9. T. Kałaczyński, M. Łukasiewicz, J. Musiał, R. Polasik, M. Szczutkowski, N. Dluhunovych, J. Wilczarska, T. Kasprowicz, Analysis of the diagnostic potential research thermovision in the technical state of combustion engine injectors assessment, Engineering Mechanics, 1805-8248 (2018)
10. M. Styp-Rekowski, E. Mańka, M. Matuszewski, M. Madej, D. Ozimina, Tribological problems in shaft hoist ropes wear process. Industrial Lubrication and Tribology, Vol. **67**, pp.47÷51 (2015)
11. R. Krystek (red), *Integrated transport safety system*, vol. I, (WKiŁ 2009)
12. Transport Research Laboratory, Construction logistics and cyclist safety, Technical report (2012)
13. Directive 206/42/WE European Parliament and Council of 17 May 2006 on unification of regulations concerning machines, altering directive 95/16/WE. OJ L 157, 26 (2006)
14. M. Balke, R. Karpiński, Road accidents due to technical reasons in statistical terms based on international and national research, Logistyka **3** (2014)
15. M. Gerigk, Formal method for safety assessment of damaged ships on the basis of analysis of causes and effects of accidents. XXXIII Winter School of Reliability
16. J.L. Frąckiewicz, *Efficient operation*, (Wrocław 1980)
17. B. Tuchańska, *Idealization and discovery. Factor and quantity,* [w:] Discovery, truth, abstraction, empiricism, history and a idealization. A. Klawiter, L. Nowak (red.), (PWN Warszawa 1979)
18. T. Szczuraszek, *Urban traffic safety*, (Editorial Office of Transport and Communication. Warszawa 2008)
19. J. Wicher, *Safety of road traffic and cars,* (Editorial Office of Transport and Communication. Issue 2, Warszawa 2004)
20. R. Madhusudana, J. Reson J, Products of Sets: Ordered and Unordered *Resonance* **21** pp 557-64 (2016)
21. D. Bąk – Gajda, J. Bąk, *Psychology of transport and road traffic safety,* (Difin, Warszawa 2010)
22. D. Meadows, *Thinking in Systems* (White River Junction, Chelsea Green Publishing 2008)
23. J. Musiał, Ł. Muślewski, M. Pająk, A. Grządziela, Analysis of vibration time histories in the time domain for propulsion systems of minesweepers. Journal of Vibroengineering **17**, 3, 1309-1316 (2015)
24. G. Dahlke, J. Kamczyc, R. Rakowski, Diagnostics and assessment of passenger car cabin ergonomics i ocena ergonomiczności, Logistic **3** (2014)
25. M Łukasiewicz, T Kałaczyński, J. Musiał, J. Shalapko, Diagnostics of buggy vehicle transmission gearbox technical state based on modal vibrations, Journal of Vibroengineering 16, 6, 3137-3145 (2014)
26. M. Matuszewski, T. Mikołajczyk, D. Yu. Pimenov, M. Styp-Rekowski, Influence of structure isotropy of machined surface on the wear process International Journal of Advanced Manufacturing Technology: Vol. **88**, Iss. 9, pp. 2477÷2483 (2017)

# Principles of Formation of a Vehicle's Remote Diagnostics

*Łukasz* Muślewski[1,*], *Roman* Zinko[2], *Oleg* Polishchuk[3]

[1]Faculty of Mechanical Engineering UTP in Bydgoszcz, Department of Vehicle Engineering, Al. Prof. S.Kaliskiego 7, 85-796 Bydgoszcz, Poland
[2]Lviv Polytechnik National University ,St.Bandery Street, 12, Lviv, 79013, Ukraine
[3]Khmelnitsky National University, Instytutska Street, 11, Khmelnytskyi, 29000, Ukraine

**Abstract.** Complex of remote diagnostics and control of an automobile's nodes enables to remotely control the state of vehicle's nodes and diagnose malfunctions; control ecological norms and compliance with traffic regulations. Its implementation will allow diagnostics of intermittent malfunctions due to application of the proposed methods of remote diagnostics of an automobile's nodes as well as to reduce terms of diagnostics significantly due to application of intellectual system of malfunctions detection and due to possibility to control parameters in real time mode. Implementation of the complex will cut unforeseen idle time of vehicles due to preventive control of the tendencies in nodes parameters changes, and due to executing modern repair or replacement of nodes and units. The complex grants safe local and remote access to data and reports by using WebPortal and VPN technologies with possibility to appoint personal right to every user. Due to application of the elaborated methods and the complex that implements it, one can achieve high rate of economic efficiency due to collection of statistic information under conditions of actual exploitation, diagnostics of emerging malfunctions of various difficulty degrees, and making necessary corrections in the structure of an automobile's nodes prior to serial production launch.

## 1 Introduction

Recently, the performance of vehicles, its power, economic, environmental and reliability indicators have increased significantly. The accelerated growth in the use of motor vehicles causes a further increase in the costs of their maintenance and repair. In this regard, the role of operational and accurate diagnostics in reducing operating costs will only increase.

The development of modern systems of several key automobile's nodes increases road safety, technical specifications and environmental safety standards. However, when complex modern car systems work together, a malfunction in one of the nodes can cause a whole complex of incorrect readings of conjugated electronic systems. Many malfunctions can manifest themselves in a time lag (the so-called alternative malfunctions). Effective diagnostics of such malfunctions is possible only in the active phase of malfunction manifestation, and is not always possible to be achieved by force. This greatly complicates the process of troubleshooting and calls for development and application of specialized methods and diagnostic tools.

Currently, most vehicle breakdowns occur on highways or city streets far from any service station. For the prevention of breakdowns, it is advisable to use remote diagnostic systems. Such systems make it possible to know information about the state of the functional nodes of a car in real time, thereby assessing its technical state.

* Corresponding author: Lukasz.Muslewski@utp.edu.pl

## 2 Problem statement

Nowadays, two main trends in the development of automobile diagnostics are distinguished. The first trend is the location of technical diagnostic tools on the automobile itself. Means of technical diagnostics can be divided into three types according to their interaction with the object of diagnosis: built-in (included in the structure), external (not included in the structure), and installed on the automobile [1].

Diagnosis at service stations does not solve the problem of restoring the vehicle's serviceability, since its main task (to quickly (if necessary) service the car) is rather complex in this case.

In addition, the use of stationary and portable diagnostics tools is usually associated with the operations of connecting, adjusting and removing sensors and connected accessories, which leads to large labor costs for supplementary work. On average, they make up from 80 to 85% of the time of the total amount of work on the diagnostics.

These drawbacks are largely eliminated by the introduction of an integrated diagnostic system. Built-in diagnostics can be implemented in two ways: by installing special control sensors built into various vehicle systems with signals output to control points; by equipping a car with sensors and additional measuring instruments that provide continuous information on the technical state of the systems, components and units of the car.

The second trend is remote diagnostics. Car manufacturing companies started to equip their cars with Internet access points, which, along with navigation, allowed to enter in the database addresses of specific dealerships where this car can be serviced. Clients have a possibility of prompting for an emergency notification, when it is time for the next scheduled maintenance; they do not need to look for the dealer's coordinates on the network, they just start navigation that shows the location of the nearest authorized center.

After the client's call, the diagnostician can immediately make a request from his workplace to the car and get information about error codes, view parameters and even start activation in order to prepare for submitting the car to a car service.

## 3 Analysis of recent research and publications

Chevrolet has launched a system called "Proactive Alerts" on some of its vehicles [2]. This system scans all nodes of the car at startup and, when errors are detected, shows notifications about them on the driver's multimedia screen. However, the main difference between the system and Chevrolet is that, theoretically, with certain special settings, it can quickly send an e-mail message about possible errors, which is already enough for technical services to monitor the state of the car in real-time. So far, it can only be installed on the Silverado, Tahoe, Suburban and Corvette models of the last year range. The company does not deny further development of this system, its distribution and implementation in a wider range of models.

Scania experimentally installs the "Scania Remote Diagnostics" system [3] on tractors. Also, "Scania Communicator", official licensed services, allow specialists to examine all the inside components of the vehivle and pre-order the required spare parts, which will significantly reduce maintenance time, and, respectively, the time saved can be used more efficiently. The related ScaniaAssistance service allows you to track and promptly call qualified specialists in case of such breakdowns when the car cannot move independently.

The main idea of remote diagnostics is to integrate telematic systems into existing information infrastructures (cellular or radio) [4]. Currently, there are a number of systems for remote monitoring of hardware using telematics equipment. These systems are used to control vehicles in various fleets, as well as for single vehicles that meet the stated requirements. Currently, the following systems operate successfully: JDLink system, USA, DEERE & COMPANY – "Remote optimization of vehicles work and logistics" and SCANIA FLEX system, Sweden, Scania AB – "Personal flexible maintenance plan" [5-8].

These systems have a wide range of capabilities that allow remote diagnostics of vehicles, but these systems are only compatible with vehicles produced by these concerns.

As a remote diagnostics, the Onstar service functionality from GeneralMotors is used, which are independent means of diagnostics, but are still used as supplementary equipment along with tools from the manufacturer.

Daimler (dealer auto scanner) offers the "Virtual Assistant" service, which is a system for transmitting data about the state of the car and remote diagnostics in real time.

## 4 Presenting mainmaterial

The diagnostics complex (Fig. 1) receives signals from the vehicle that display its technical state, namely, operating characteristics, registration data, and its functional nodes. The complex recognizes the incoming data, monitors the level of performance degradation, detects probable malfunctions, and sends a signal to the vehicle with an assessment of the technical state of the vehicle and its functional nodes, as well as recommendations to adjust the characteristics to optimal ones. Information is transmitted between the vehicle and the complex by means of telecommunication. The signals transmitting operational characteristics of the car and its functional nodes are received from the controllers through the test connector of the car.

Companies that are interested in the development of car maintenance create a technical support service to work with clients via the Internet. They create a websites containing software libraries and databases on technical specifications, troubleshooting, availability of spare parts and addresses of repair enterprises, etc.

Car owners purchase an adapter that connects the test connector to the modem, a cell phone (smartphone) and / or additionally connect the adapter to the laptop. Car owners are registered in an automated diagnostic system. They receive the necessary software for their car from the technical support service.

Car owners perform diagnostics using a local system in the car; if necessary, go online and select the type of service.

With the help of the software using the database available on the server, car diagnostics is performed and appropriate instructions are given to the car owner.

For the lower level of this system (a local automatic vehicle malfunction diagnostics system), a design is being developed in the form of an interface unit connected to a computer. Programs for the microcontroller of the interface unit and computer have been developed.

In most modern cars, there are multiplex data buses, through which data is exchanged with various car self-diagnostics systems. There may be several of them, different in speed, protocol and set of transmitted data. The configuration and tire protocols depend on the level of development of the car's self-diagnostics systems.

Most communication protocols (CAN, Lin, DeviceNet, CANopen, SAE J1939) [9-14] and adapters are supported by the WarwickControlTechnologies X-Analyser software. When using free version one can see information transmitted via the bus on the display in numerical form.

An example of a multiplex data bus is the CAN bus. CAN communication protocol is standardized according to ISO 11898-1 (2003). This standard mainly describes a data exchange layer consisting of a logical control (LLC) subsection and an access control (MAC) subsection, and some aspects of the physical layer of the ISO / OSI model. The remaining layers of the protocol are left to the discretion of the network developer [15]. CAN benefits:

- affordability;
- protocol realization on hardware level;
- primitive transmission line;

- superb ability to detect errors and malfunctions as well as their localization;
- malfunctions detection and check system.

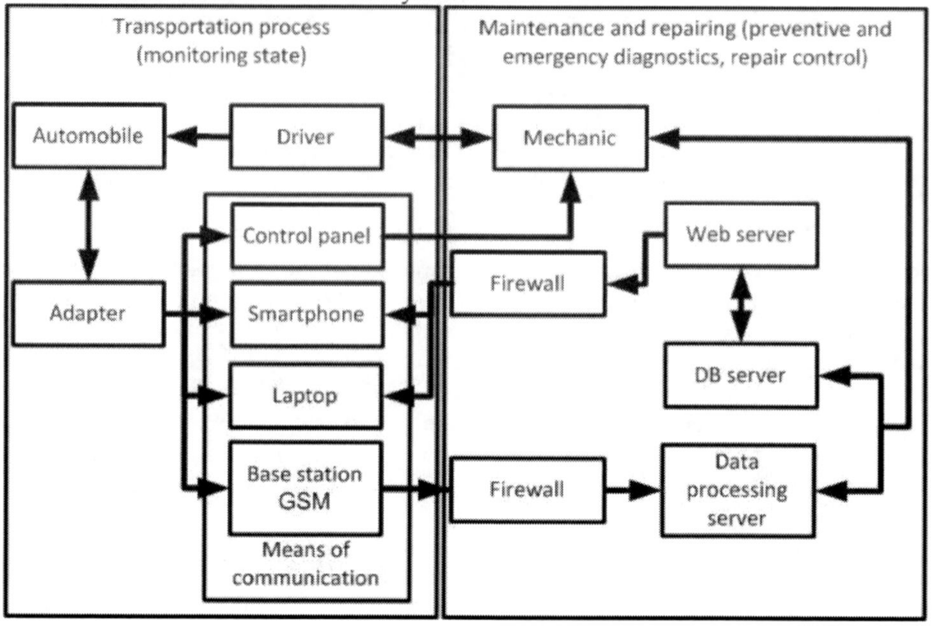


**Fig. 1.** Complex of remote diagnostics and control of the state of an automobile's nodes.

CAN is a system based on a serial bus with multifunctional capabilities; all CAN nodes are capable of transmitting data, and some CAN nodes can request a bus at the same time. The transmitter transmits a message to all CAN nodes. Each node, based on the identifier obtained, determines whether it should process the message or not. The identifier also determines the priority that the message has when accessing the bus [9].

Modern cars have electronic systems, operation of which is based on the parameters of the car controlled by a set of sensors on various nodes that can signal an error in their work with a specific diagnostic code of the self-diagnostics system. When several such systems interact, a malfunction of one of the parameters can cause incorrect operation of several systems at the same time. At the moment of error, each of the systems signals the code of the alleged cause. Error diagnostics can be greatly reduced by obtaining important information from the relevant codes. Diagnostic standards have been developed to unify the interaction protocols and messages format of these systems.

On-board navigation equipment (Fig. 2) interacts with the main technological components of the technical state monitoring system of the vehicle within the framework of the ideology of a mobile information-diagnostics system. The signals from sensors and vehicle systems are processed, and then an array of messages is generated. To obtain information from the diagnostics control systems that are part of the nodes and units of the vehicle, the information-diagnostic system generates a series of necessary commands. Having received encoded information from the standard built-in control systems, the information-diagnostic system, using its own software, determines the codes and creates an array of diagnostic messages. The purpose of identifying codes is to recognize vehicle states that are dangerous for continuing further work, first of all, creating a risk of expensive repairs, accident, loss of a dealership guarantee and other unfortunate consequences.

When a malfunction occurs during movement, it calls for recognition of the signs of malfunctions and then makes various decisions about the possibility of continuing the movement. Diagnostics information received from the diagnostics port or from the sensors of the standard on-board diagnostics system is processed on board in an automated mode [16].

Diagnostics system with data processing software is the most flexible. It allows you to read OBD-II codes and data streams in real time and present them in an intuitively clear format, in numerical form, in a form of description of possible malfunctions, in a form of tables, as well as in graphical form, including a form of multiparameter graphs. Using such system, you can conduct virtual tests: manually change one of the parameters and see what happens to the rest. At the same time, the protocol necessary for a detailed analysis of transients is done in real time. Such protocols can be useful for routine diagnostics: you can gradually accumulate an "engine history" and identify probable problems in a timely manner. Such diagnostics system can be designed on the basis of wireless technologies for data transmission and remote analysis in a data center.

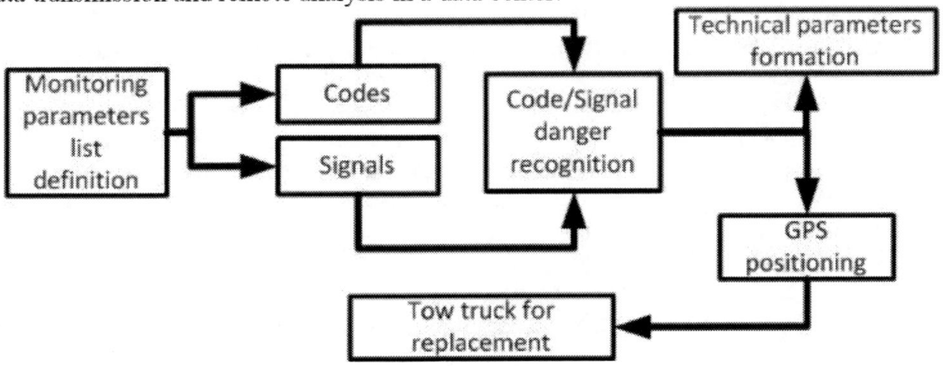


**Fig. 2.** Algorithm of vehicle's technical state monitoring system.

Currently, the most common wireless data transfer technologies are as follows:

- GPRS, EDGE (GSM standard);
- CDMA;
- WiFi.

More modern HSDPA technologies, WiMAX which are 3.5G-4G generation standards are in test operation. When comparing various wireless access technologies, the following points should be taken into account. First of all, the higher the frequency range, the smaller the communication range (cell radius) is, but the higher the data rate is. In this regard, for example, from the point of view of lowering capital expenditures in areas with low population density, it is preferable to build CDMA2000 networks in the 450 MHz range, and to compare broadband wireless access technologies of large cities, CDMA lx EV-DO networks in more high-frequency ranges (1.9 or 2.1 GHz) [17].

Existing diagnostics algorithms have an insufficient diagnostics level when analyzing the causes of complex and "intermittent" malfunctions, due to the lack of the possibility of applying real-time diagnostic techniques and the possibility of executing diagnostic actions at the main nodes, instantly receiving a system response to this effect exactly when the error is in the active phase.

To eliminate these shortcomings, we propose a method for remote diagnostics and monitoring of the state of an automobile's node (Fig. 3). A method for remote diagnostics consists in collecting diagnostic parameters at the monitoring object, initial processing and buffering, transfer of data using modern wireless technologies to a remote data processing server, and further analyzing and predicting technical state of the vehicle.

The method of remote diagnostics is described by remote diagnostics algorithms. Due to technical complexity of the systems of modern vehicles and the interpenetrating interaction of its nodes, there are malfunctions, the active phase of which is manifested by chance.

Therefore, such intermittent malfunctions may not occur during local diagnostics directly at the service center. Diagnostics and determination of the cause of these malfunctions becomes impossible in the passive phase of the malfunction. Very often, all nodes of the vehicle work exclusively in normal modes and diagnostics by any method or equipment will not show any deviations. However, the malfunction will stay and may reappear when a combination of various external and internal factors cause it. The method of remote diagnostics allows remote monitoring of the safety of vehicles, as well as to fix the manifestation of the active phase of the malfunction and obtain important information during various diagnostics actions, and determine the true source of the malfunction.

One of the cyclic processes is the collection of sensor readings, primary diagnostics, formation of a data packet to transmit and forward information. This process can be configured based on the set priorities of the parameters of the collected information. This is necessary for the packet for transmission to be optimal, carry all the basic information, and do not contain redundant information. Based on the specifics of the application, a personal priority can be assigned to each of the transmitted parameters, which will affect the transmission of the parameter by means of wireless connection. On the part of the data center, the process of receiving, storing and analyzing information from diagnostic objects is cyclically active. The database is being constantly updated with the transmitted parameters, on the basis of which it is possible to obtain interconnected diagrams and trends in the vehicle parameters. Upon occurrence of an error, all basic information, if necessary, will be provided to the operator for further processing [6]. Having remote access to the information provided by the data processing server, a diagnostic engineer can:

- track the readings of car sensors in real time (on-line);
- receive current or previously recorded error codes from the self-diagnostics systems;
- conduct remote diagnostics based on the OBD-II standard (if this is supported by the vehicle's systems);
- receive data from sensors in diagnostics mode, for the analysis of transient processes with a frequency of 50-500 Hz;
- get information on typical situations similar to the situations already recorded in the intelligent system to detect malfunctions;
- get information about the dynamics of changes in the parameters of the nodes and get the response of the system nodes when executing various diagnostics actions, immediately in the active phase of the manifestation of an intermittent malfuntion.

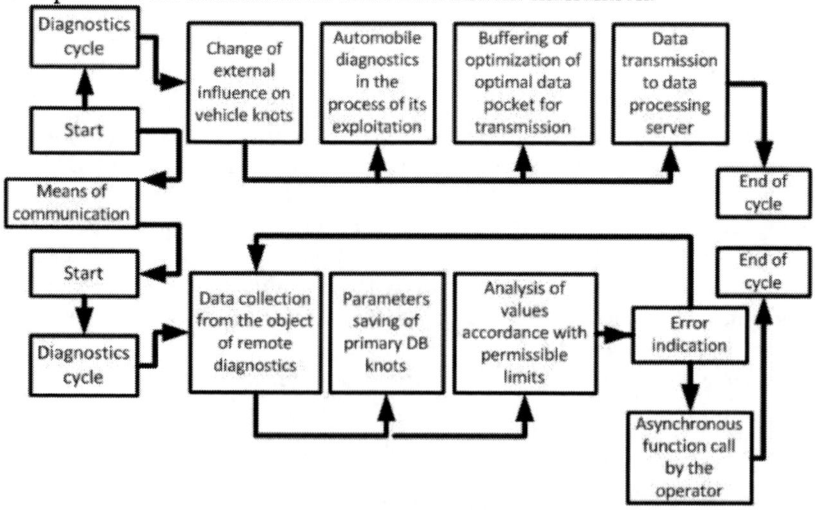


**Fig. 3.** Functional scheme of operation of assembly modules, receiving and transmission, data processing.

The complex of remote diagnostics and control of the state of vehicle nodes implements the remote diagnostics technique, which enlists collection of diagnostics parameters at the object of control, initial processing and buffering, transfer of data via modern wireless technologies to a remote data processing server and further analysis by means of an intelligent error detection system, graphical analysis by various interpolation algorithms.

The hardware and software complex includes:

- the hardware diagnostics module, which is a device installed on the car and switched to the multiplex data bus through a test connector;
- data processing server, which is a service center that receives, stores and processes data from objects of control;
- software for processing data from objects of control, their analysis and assessment of the safe operation of vehicles on the basis of an intelligent system of malfunctions detection;
- prediction of vehicle's state, which defines the terms of malfunctions [18].

The complex has a modular structure that allows you to connect high-speed modules for wireless data transmission, based on modern EDGE, CDMA technologies etc., allowing you to implement a completely remote process for diagnostics of the state of the automobile's nodes. At the object of control, the operation modes of the complex can be observed on LEDs that reflect the state of the system (monitoring, diagnostics, malfunction). In the complex, control of the state of the automobile's nodes is carried out by control of the basic parameters of the node and analysis of diagnostic codes received from self-diagnostics systems. To solve the problems of managing the complex and providing a convenient interface for working with the received data, ESI tronic software can be used.

The software of the remote diagnostics complex can function under the control of Windows, Linux operating systems etc. The requirements for the hardware configuration of a computer are determined by the requirements of the operating system.

Prediction of the state of the vehicle is based on four types of initial information: statistics of repairs, expert assessment of the resource, statistics of diagnostics, data on technological loads. In this case, a diagnostics model is used.

Diagnostics parameter-forming model [19, 20] is a mathematical apparatus that uses data on the timing and results of the diagnostics performed in the process of its functioning. The peculiarity of this model is obtaining of the value of the residual resource of the unit based on data on the current technical state. Totality of output data for this model is presented as "Diagnostics statistics". This model can be represented by the formula (1):

$$[Z,M,v] = f[Y(t_i)S(Y,t_i)] \tag{1}$$

in which $Y(t_i)$ - value of the diagnosed parameter $Y_i$ for the exploited unit (current implementation) at the moment $t_i$;

$S(Y,t_i)$ - diagnostics data statistics on previous implementations.

In this model, only $t_i$ value is determined; it is described by the distribution law of Z, M and v. The starting point of $t_y$ forecast forestalling is the time of the last diagnostics.

## 6 Conclusions

The complex of remote diagnostics and control of the state of an automobile's nodes will enable remote control of the state of vehicle node and troubleshooting, control compliance with environmental standards and traffic regulations. Its implementation will enable diagnostics of intermittent malfunctions due to application of the proposed method of remote diagnostics of an automobile's nodes, as well as significantly reduce time for diagnostics due to use of an intelligent system for detection of errors and the ability to control parameters in real time. Implementation of the complex will reduce unforeseen idle of vehicles due to preventive control of trends in the nodes parameters, and timely repair or replacement of nodes and units. The complex implements secure local and remote access to data and reports using WebPortal and VPN technologies, with the possibility to assign personal rights to each user.

Due to application of the elaborated methods and the complex that implements it, one can achieve high rate of economic efficiency due to collection of statistic information under conditions of actual exploitation, diagnostics of emerging malfunctions of various difficulty degrees, and making necessary corrections in the structure of an automobile's nodes prior to serial production launch.

## References

1. The basic principles of diagnosing MSHT using modern diagnostic equipment / P.S. Sinitsin, H.D. Kokoriev, I.A. Uspenskyi // Collection of scientific works of students of the Ryazan State Agrotechnological University named after P.A. Kostychova. – Ryazan, – P. 263-269, (2011)
2. Proactive Alerts analyzes the condition of the car: https://petrenco.com/avto.php?txt=605/ (date of the application: 03.11.2019)
3. Extention useful life – with technology Scania Remote Diagnostics http://scanauto.ru/media-center-scania/news-scania/uvelichenie-vremeni-poleznogo-ispolzovaniyaavtomobilya-pri-pomowi-sistemy-udalennoj-diagnostiki-scania-remote-diagnostics. (date of the application: 03.11.2019)
4. Fuks V.A. Universal system for remote vehicle diagnostics // Young scientist. – 2019. – №12. – P. 40-44. – URL https://moluch.ru/archive/250/57481/ (date of the application: 03.11.2019)
5. Denton, T. (2012). Automobile electrical and electronic systems: Automotive technology: Vehicle maintenance and repair. NewYork: Routledge, (2012)
6. Pupkov K.A., Konkov V.G. Intelligent systems: research and creation. - M.: Publishing House of MSTU named after N.E. Bauman, 2003. – 345 p., (2003)
7. Website www.deere.com – company official website John Deere. (date of the application: 03.11.2019)
8. Website www.scania.com LLC «Scania Service " – The official dealer of the Swedish automotive concern Scania CV AB in Russia. (date of the application: 03.11.2019)
9. http://www.can-cia.org/ – Official website of an international organization CAN In Automation. (date of the application: 03.11.2019)
10. http://www.obdii.com/ – Official website of the international standardization organization OBD II. (date of the application: 03.11.2019)
11. Dzhelekarski P., Alexiev D. Initializing Communication to Vehicle OBDII System. Proc. Intern. Conf. Electronics, 2005, vol. 3, pp. 46–52., (2005)
12. Malekian R., Moloisane N.R., Nair L., Maharaj B.T., ChudeOkonkwo U.A.K. Design and Implementation of a Wireless OBD II Fleet Management System. IEEE Sensors J., 2017, vol. 17, no. 4, pp. 1154–1164, (2017)
13. Zhang H., Kang W. Design of the Data Acquisition System Based on STM32. Procedia Comp. Sc., 2013, vol. 17, pp. 222–228, (2013)
14. Si H., Aung Z.M. Position Data Acquisition from NMEA Protocol of Global Positioning System. IJCEE, 2011, vol. 3, no. 3, pp. 353–357, (2013)
15. Garanin M.V. Systems and networks of information transfer. – M.: Radio and communication, 2001-336 p., (2001)

16. Principles of developing a telematic system for monitoring the technical condition of cars / S. V. Zhankaziev, S. P. Ignatiev // Bulletin of the Moscow Automobile and Road State Technical University (MADI). – 2011. – № 3. – P. 22-28, (2011)
17. Pljetnjev S.V. Engine of truck diagnostic on the basis of technology ODX / S.V. Pljetnjev, J.V. Krjukov, A.V. Ferenc // Announcer of the Kazan state technical university the name of A.N. Tupoljev. – 2014. – № 2, – P. 58–61.
18. Shakhnovich I.V. Modern wireless technology. – M.: Technosphere, 2006. – 288 p., (2014)
19. Aubekerov N.A. Individual assessment of operational damage to vehicle parts / N.A. Aubekerov, J.N. Aubekerova, A.N. Abatova // Automotive industry. – 2014. – № 11. – P. 18–20., (2014)
20. M. Łukasiewicz, T. Kałaczyński, J. Musiał, J. Shalapko. Diagnostics of buggy vehicle transmission gearbox technical state based on modal vibrations, Journal of Vibroengineering, **16**, 6, (2014)

# Construction proposal of hydraulic equipment designed to test hydraulic components and energy carriers

*Jozef* Nosian[1,*], *Marietta* Markiewicz[2] , *Romana* Janoušková[3] and *Patrícia* Feriancová[3]

[1]Slovak University of Agriculture, Department of Machine Design, 94976 Nitra, Slovakia
[2]UTP University of Science and Technology, Faculty of Mechanical Engineering, 85-796, Bydgoszcz, Poland
[3]Slovak University of Agriculture, Department Transport and Handling, 94976 Nitra, Slovakia

**Abstract.** The aim of the present article is the construction of a device for measuring and testing hydrostatic converters and subsequent verification measurement of the proposed device. This device will allow the measurement and testing of hydraulic pumps under laboratory conditions. In our case, we will test the hydraulic pump UD–25R, which is used in hydraulic drives of agricultural machines. Last but not least, the device described in the article can be used to test individual energy carriers used in hydraulic systems. The described device can simulate conditions that occur in operation. Using verification measurements we have proved that the designed laboratory equipment is suitable for testing hydraulic components and also allows monitoring and testing of the properties of individual hydraulic fluids. The measurement will result in a change in flow rate over time.

## 1 Introduction

Hydraulic devices are widely used in all areas of industry, mainly for their advantages. The biggest benefits of hydraulic mechanisms include: transmission of energy over long distances, high performance, simplicity, reliability and a relatively long life. Current development and modernization in the field of hydraulic mechanisms is mainly aimed at reducing energy consumption and increasing the power that these machines are capable of transmitting. Last but not least, the development of these mechanisms focuses on their environmentally friendly operation, ie that these mechanisms will minimize contamination of the environment. It is very important that hydraulic components and energy carriers are tested under laboratory conditions before being put into practice. There are high requirements of hydraulic fluids that serve as energy carriers in hydraulic systems. Their producers must consider their demands for quality improvement and simultaneously reducing the burdening of environment [1]. A comprehensive analysis of the effect of biological fluids and mixtures thereof has already been performed by Puškár et al. [2]. Before their first use in operation, fluids must be subjected to laboratory tests to prevent or minimize risk of damage of expensive agriculture machinery. A specially designed laboratory device observing the operating conditions of machine is used for this purpose [3]. The need for agricultural machinery to be tested from the point of view of their suitability to agricultural use will grow continuously because these machines directly affect agricultural production [4]. The measurement of hydraulic devices used in industry was also dealt by authors Simikič et al. , Zastempowski and Hujo et al. [5, 6, 7].

## 2 Material and methods

* Corresponding author: xnosian@uniag.sk

### 2.1 The determination of the individual components of a hydraulic device

The determination of the individual components of a hydraulic device for assessing the operation of the hydraulic pumps and energy carries are based on relations from author Hujo et al. [3].

Wall thickness of a pipe:

$$s = \frac{p_{max}.d_p}{2.\delta}.k \tag{1}$$

where:
$p_{max}$ – max. pressure, MPa
$d_p$ – inner surface, $m^2$
$\delta$ – ultimate tensile strength, Pa

$$d_p = \sqrt{\frac{4.Q}{\pi.w}} \tag{2}$$

where:
$Q$ – flow rate, $dm^3$.rpm
w – resistance to flow rate, m/s
Cooling power:

$$P_0 = P_{01}.1.1 \tag{3}$$

Specific cooling power:

$$P_{01} = \frac{P_V}{T_1 - T_2} \tag{4}$$

where:
$P_v$ – cooling power, W
$T_1$ – max. temperature, °C
$T_2$ – min. temperature, °C

Based on the presented calculations we were able to determine the basic parameters of the experimental laboratory equipment. The calculated values are shown in Table 1. We were able to determine further necessary data on the basis of calculations. Based on the data from the catalog, we were able to select other required components with the characteristics and parameters necessary for commissioning the laboratory equipment (Table 1).

**Table 1.** The basic parameters of experimental laboratory equipment.

| Parameter | Value |
|---|---|
| Oven wall thickness (min.) | 1.21 mm |
| Inner surface (min.) | 9.86 $mm^2$ |
| Cooling capacity (min.) | 0.1215 kW/°C |
| Specific Cooling Performance (min) | 0.1165 kW/°C |
| Pipe diameter (d) | 14 mm |
| Wall Thickness (s) | 1.5 mm |
| Tank Capacity (V) | 80 $dm^3$ |

When operating an experimental device for measuring and diagnosing the operation of hydraulic pumps and hydraulic fluids, heat is formed in the hydraulic circuit. This is why a cooler is installed on the device. Cooler is used to heat dissipation the hydraulic circuit. In our case it is an EMMEGI cooler type 20124 K 230–400. We chose this cooler based on calculations. The cooler performance is $P_0$ = 0.2 kW/°C. According to the above calculations, this power is sufficient to cool the hydraulic circuit. The evaluation of hydraulic fluids in dependence on temperature and pressure was dealt with by Kopiláková et al. [8] and Kosiba et al. [9]. Both authors found that the operating parameters, mainly temperature and pressure, affect the working fluid and affect subsequent measurements. On Figure 1 it is possible to see a diagram of an experimental hydraulic device for measuring and testing hydraulic pumps and energy carries. The equipment can be used to measure and test various hydraulic pumps and hydraulic fluids. Figure 1 shows a device for measuring and testing hydraulic pumps and energy carriers. From Figure 1 we see that the device consists of two circuits, namely circuit A (left side) and circuit B (right side). These two circuits are separated by a mechanical coupling (MC). Circuit A and circuit B practically consist of the same components to simulate the same conditions in both circuits. One part of the circuit is used to test and examination a hydraulic pump with a tilting plate (circuit A). The second part of the circuit is used to test and diagnose the rotary gear pump. This gear pump has external toothing (circuit B). Each circuit has a separate tank. This makes it possible to test two different types of hydraulic fluids. Filters are placed on the tanks. The filters are used to filter out dirt from hydraulic oil. As in any hydraulic circuit, a safety valve must also be fitted in this circuit. Each circuit has its own safety valve, in our case both safety valves are set to maximum possible pressure of 20 MPa. After exceeding this pressure value, the safety valve releases the oil from the hydraulic circuit back into the tank. UD 25 is a one-way gear pump from Jihostroj Aero Technology and HydraulicsUD-25R is most commonly used in agriculture, specifically in ZETOR tractors and most often in a ZETOR Forterra tractor. The technical parameters of the hydraulic pump are given in Table 2.

**Table 2.** The basic parameters of experimental laboratory equipment.

| Parameter | Unit | Value |
|---|---|---|
| Rated speed | $min^{-1}$ | 1500 |
| Maximum speed | | 3000 |
| Minimum speed | | 500 |
| Maximal pressure at the inlet | MPa | 0.05 |
| Minimum inlet pressure | | 0.03 |
| Nominal outlet pressure | MPa | 20 |
| Maximum outlet pressure | | 23 |
| Geometric volume | $cm^3$ | 25 |
| Maximum oil viscosity | $mm^2.s^{-1}$ | 1200 |
| Minimum oil viscosity | | 10 |
| Maximum oil temperature | °C | 80 |
| Minimum oil temperature | | -20 |

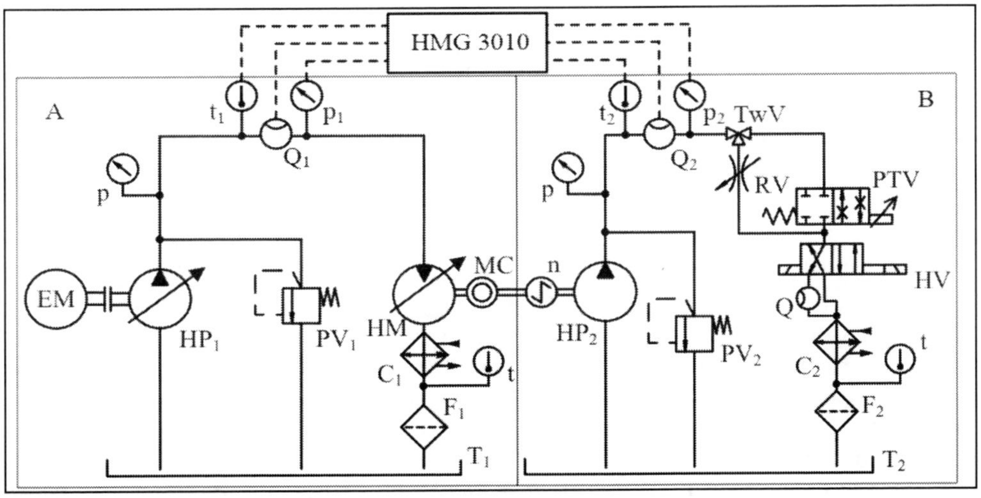


**Fig. 1.** Kinematic scheme of hydraulic circuit. Legend: (A – primary circuit, B – secondary circuit, EM – electromotor, $HP_1$ – regulatory hydraulic pump, $PV_1$, $PV_2$ – pressure valve, p – manometer, $t_1$, $t_2$ – temperature sensor ETS 4148-H-006-000, $Q_1$, $Q_2$ – flow sensor EVS 3108-H-0300-000, $p_1$, $p_2$ – pressure sensor HDA 4748-H-0400-000, Q – volume flowmeter, t – temperature gauge, HM – regulatory hydromotor, $HP_2$ – gear hydraulic pump, RV – reducing valve, HV – hydraulic valve, PTV – proportional throttle valve, TwV – three-way valve, MC – mechanical coupling, n – rpm sensor ISP 308 S 01 N, $F_1$, $F_2$ – filter, $T_1$, $T_2$ – tank).

Part A is used to drive and control circuit B. EM is an electric motor that is used to convert electrical energy into mechanical energy. This electric motor replaces a combustion engine, which is commonly used in hydraulic pump drive machines. The mechanical energy with the electric motor is converted by the hydraulic pump $HP_1$ into the pressure energy in the hydraulic circuit. At the top we have the HMG 3010, which is used to monitor the values. In our case, the device monitors temperature, pressure, and flow. It records and stores the monitored values in its memory. The HMG 3010 is connected to pressure, temperature and flow sensors. In both circuits we have installed safety valves $PV_1$ and $PV_2$ and filters $F_1$ and $F_2$. As already mentioned, heat is generated when passing the liquid through the circuit. This heat is an undesirable effect and therefore both circuits need to be cooled using $C_1$ and $C_2$ coolers. The cooler maintains a constant temperature of $t = 40$ °C. Dilatation thermometers are located in both tanks to monitor the hydraulic oil temperature. Both hydraulic circuits have their own tank $T_1$ and $T_2$. So we can have two types of oils in the device at the same time. One type of oil in one tank and another type of oil in the other tank. The HM hydraulic motor is used to convert the pressure energy generated by the $HP_1$ hydraulic pump into mechanical energy. The hydraulic motor is connected to the $HP_2$ hydraulic pump via the MC coupling. The $HP_2$ hydraulic pump is a gear pump and converts mechanical energy into pressure energy in the B circuit. This part of the circuit is largely identical to circuit A. PTV is used to increase the hydraulic pump load. Using a proportional throttle valve, we can simulate the pressures that are achieved when operating a machine used in practice. The TwV three-way valve is used to change the selected hydraulic pump load method. When the reduce valve is used, it changes the pump flow and also changes the pressure in the hydraulic circuit. The flow change process can be controlled by a solenoid, using a PTV proportional throttle valve. A second alternative that serves to verify the flow characteristics of a hydraulic pump is to use a volume flow meter. Flow control through the flow meter is controlled by a two-position hydraulic valve HV. The two-position switch switches to the working fluid flow sensing position and is measured exactly $t = 60$ s. After the measurement, the flow characteristic of the hydraulic pump is obtained. Thanks to this characteristic we are able to verify the functionality of the hydraulic pump connected in the laboratory equipment. The circuit also includes manometers (P), pressure sensors ($p_1$ and $p_2$) and temperature sensors ($t_1$ and $t_2$) for temperature and pressure monitoring. The laboratory testing device allowed to simulate variable testing condition to simulate the real agricultural wheel tractor operational condition. The flow properties of the hydraulic pump with tested hydraulic fluid [10]. The measurement results in a change in flow and pressure on the time.

## 3 Results

### 3.1 Laboratory measurements

The laboratory equipment consists of two separate circuits. Proportional (PTV) resp. the throttle valve (TrV) is used to exerting a load in the circuit. This load simulates operating pressures. Thrust valve loads have been used by their authors Hujo et al. [3] and Kosiba et al. [11] in their laboratory hydraulic devices. Thanks to the construction, we can also install another type of hydraulic pump into the circuit. The verification measurement was carried out in the laboratory of the Department of Transport and Handling, of the Faculty of Engineering of the Slovak University of Agriculture in Nitra. Verification measurements were made at 1 500 rpm. The output of these measurements was the flow of the hydraulic pump. Based on the measurements, we created a graph (Figure 2, Figure 3).

### 3.2 Flow rate and flow efficiency of the hydraulic pump

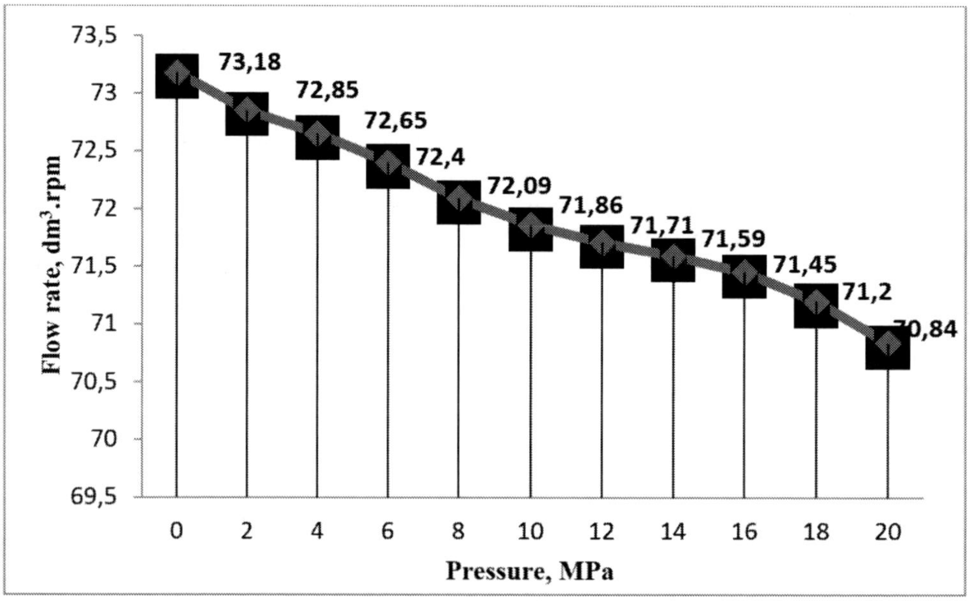


**Fig. 2.** Flow characteristic of the hydraulic control pump.

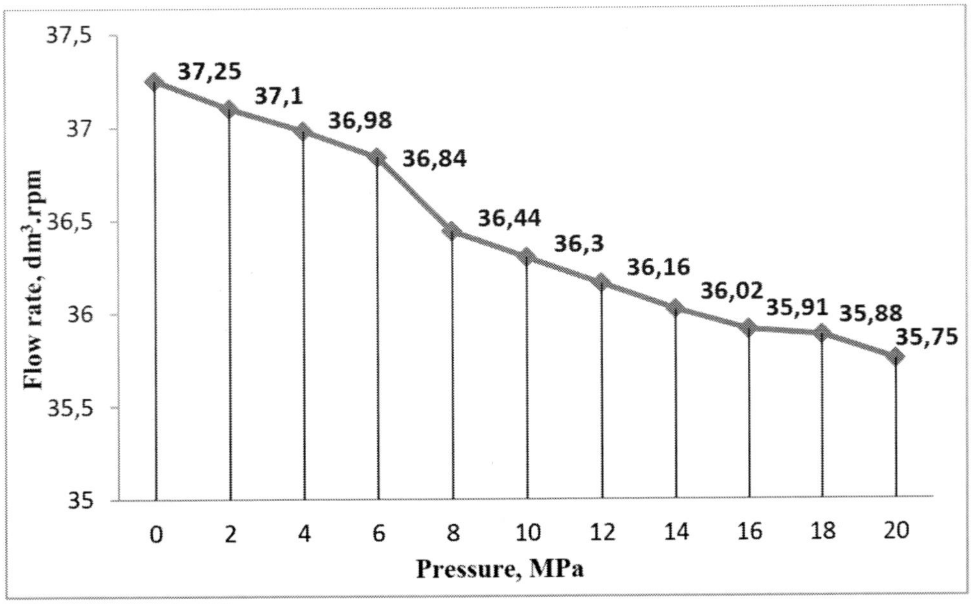


**Fig. 3.** Flow characteristic of hydraulic gear pump.

Flow efficiency of hydraulic pump (Figure 4, Figure 5) will be calculated form the measured flow characteristic:

$$\eta_p = \frac{Q}{V_G.n}.100 \tag{5}$$

where:
$V_G$ – geometrical volume of hydraulic pump, dm3
$n$ – nominal speed of hydraulic pump, rpm

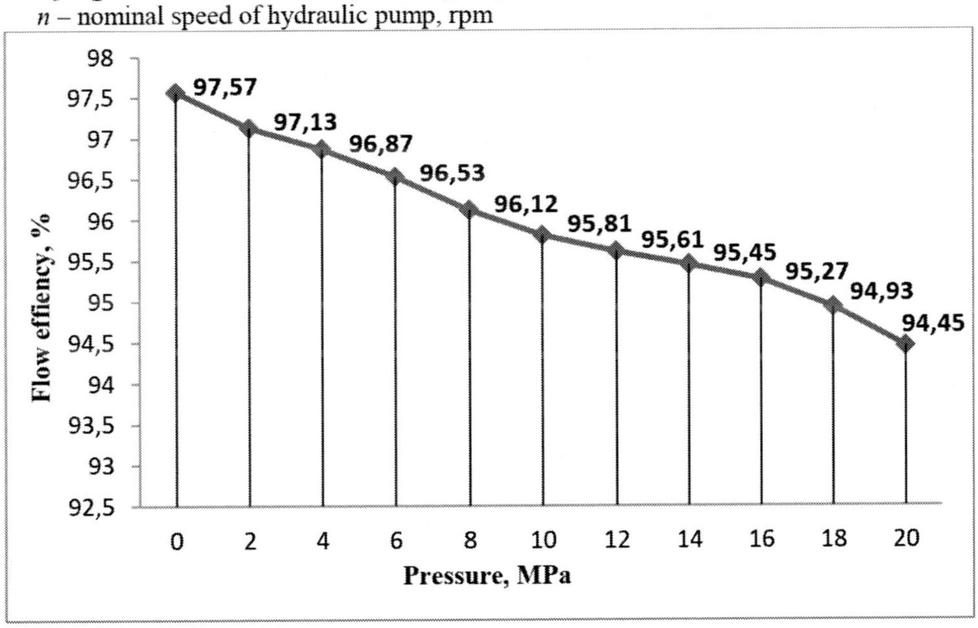


**Fig. 4.** Flow efficiency of the regulating hydraulic pump.

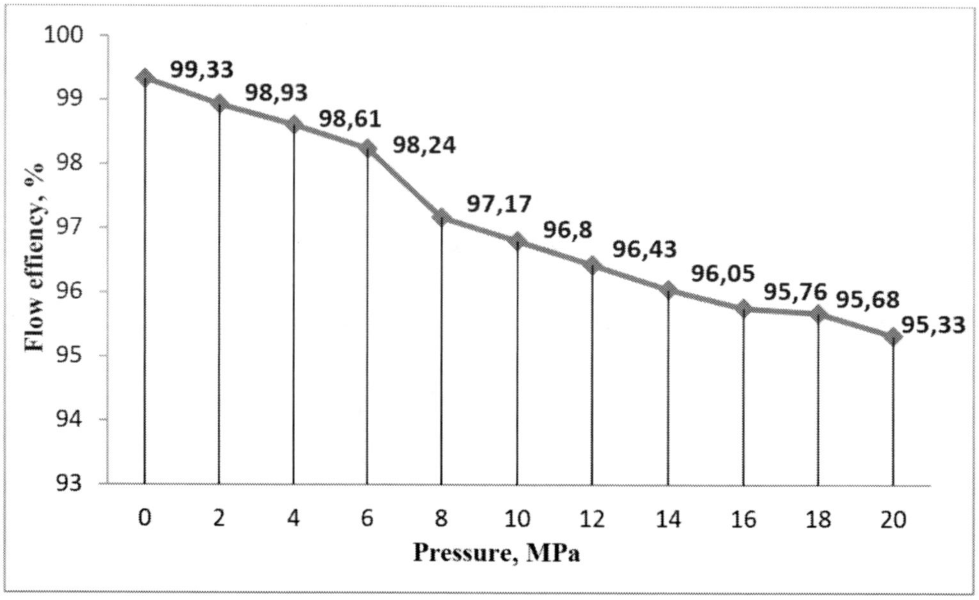


**Fig. 5.** Flow efficiency of the gear hydraulic pump.

With correct data analysis and correct interpretation of data we can create very precise results [12]. During the verification measurement, we were able to record the flow values of the gear hydraulic pump (circuit B), at a pressure p = 0 MPa and a rotation speed n = 1 500 rpm. Thanks to these values we were able to determine the flow efficiency of the hydraulic pump based on the calculations. In this case, we calculated the efficiency $\eta_{pp} = 99.33\%$. By changing the pressure during the measurement, p = 20 MPa, we calculated the flow efficiency $\eta_{pp} = 95.33\%$. In circuit A, in the regulatory pump circuit, we recorded a flow efficiency of ηpp = 97.57% at a pressure of p = 0 MPa and at a speed of n = 1500 rpm and at pressure p= 20MPa we noticed ηpp = 94.45%.

Reducing the flow efficiency of the hydraulic pump relative to the increasing pressure corresponds to the results of the work of authors Kosiba et al. [11] and Tulík et al. [13]. On the basis of the theoretical processing of the authors Hujo et al. [7] and Kosiba et al. [9] it is possible to determine the coefficient of loss of liquid $k_Q$, which affects the resulting liquid flow.

Displacement gear pumps:

$$q = 2.z_m(V_1 - V_{loss}) = 2.z_m\left[\frac{n}{360}\pi.r_{a1}^2 - \int f(x_0, y_0)dx - V_{str}\right] \quad (6)$$

where:
$z_m$ – number of tooth gaps,
$V_1$ – volume of tooth gap, $m^3$
n – rotation speed, rpm
$V_{loss}$ – loss volume , $m^3$
$r_{a1}$ – radius of the gearhead, m
Loss volume:

$$V_{str} = \int_{x_{j1}}^{x_{j2}}[f(x_2, y_2) - f(x_1, y_1)dx] \quad (7)$$

where:
$f(x_2,y_2)$ a $f(x_1,y_1)$ – gear rotation function
Resulting Loss Flow Ratio:

$$k_Q = \frac{Q_{max} - Q_{min}}{\bar{Q}} \quad (8)$$

where:
$Q_{max}$ – maximum flow of fluid, $m^3.s^{-1}$
$Q_{min}$ – minimum flow of fluid, $m^3.s^{-1}$
Resulting Loss Flow Ratio:

$$k_Q = \frac{Q_{max} - Q_{min}}{\bar{Q}} \quad (9)$$

where:
$Q_{max}$ – maximum flow of fluid, $m^3.s^{-1}$
$Q_{min}$ – minimum flow of fluid, $m^3.s^{-1}$

## 4 Conclusion

Hydraulic equipment is widely used in agricultural machinery. Thus, it is very important to design a device that would allow testing and diagnosis of hydraulic components under laboratory conditions. The described device can test and measure hydraulic pumps as well as energy carriers. Recently, more and more emphasis has been placed on ecological liquids, therefore there is a need to design a device that would allow testing of ecological liquids. In view of high ecotoxicity and low biodegradability, mineral oil-based lubricants constitute a considerable threat to the environment [14]. Therefore, it is necessary to examine the idea of the possible harmfulness of agricultural machinery on the environment [15]. The analysis of operating fluids was dealt with by authors Čorňák [16] and Tkáč [17]. The main requirement for the equipment was to simulate as much as possible the actual hydraulic circuit of the agricultural machine. The proposed device facilitates testing of hydraulic pumps and individual energy carriers used in hydraulic systems. Based on the verification measurement, we have shown that the device described in the article can be used for measuring and testing hydraulic pumps and energy carriers.

This work was supported by project VEGA 1/0155/18 „Applied research of the use of ecological energy carriers in agricultural, forestry and transport technology."
This work was supported by project KEGA 028SPU–4/2019 „Practical utilization of design and testing knowledge of transmission systems of hydraulic mechanisms of mobile agricultural and forestry machinery."
This work was supported by project APVV SK-PL-18-0041 „The development of scientific cooperation in the study of the effects of biofuels in road transport, including environmental impact."

## References

1. J. Rusnák, M. Kadnár, M. Kučera, *Biologicky odbúrateľné oleje z pohľadu ich tribologických vlastností*. SPU Nitra, 85 (2009)
2. M. Puškár, A. Jahnátek, I. Kuric, J. Kádárová, M. Kopas, M. Šoltésová, Air Qual., Atmos. Health, **12**, 7 (2019)
3. Ľ. Hujo, P. Kangalov, J. Kosiba, *Laboratory test devices for evaluating the lifetime of tractor hydraulic components*. University "Angel Kanchev" of Ruse, 69 (2015)
4. Ľ. Hujo, J. Jablonický, Z. Tkáč, *Návrh inovatívneho laboratórneho simulačného zariadenia na skúšanie hydrostatických prevodníkov a hydraulických kvapalín*. SPU Nitra. 140 (2017)
5. M. Simikič, N. Dedovič, L. Savin, M. Tomič, O. Ponjičan, Soil Till. Res., **141**, 32–43 (2014)
6. M. Zastempowski, Journal Res. Appl. in Agrc. Engineering, **58**, 2 (2013)
7. Ľ. Hujo, Z. Tkáč, J. Tulík, J. Kosiba, D. Uhrinová, M. Jánošová, RAE, **62**, 1 (2016)
8. B. Kopiláková, J. Turza, Ľ. Hujo, J. Kosiba, Tribology in industry, **39**, 129–135 (2017)
9. J. Kosiba, Š. Čorňák, J. Glos, J. Jablonický, V. Vozárová, A. Petrovič, J. Csillag, Agronomy Research, **14**, 5 (2016)

10. Ľ. Hujo, Š. Čorňák, Z. Tkáč, M. Jánošová, In 7th Intern. conf. on trends agric. engineering. Prague, Czech Republic, 183–188 (2019)
11. J. Kosiba, Z. Tkáč, Ľ. Hujo, B. Stančík, I. Štulajter, RAE, **59**, 27–33 (2013)
12. I. Kuric, I. Zajačko, M. Císar, Adv. in Scie. and Techn. Research journal, **10**, 32 (2016)
13. J. Tulík, Ľ. Hujo, B. Stančík, P. Ševčík, Journal of Central Euro. Agric., **14**, 4 (2013)
14. M. Kučera, Z. Aleš, Z. Ivandič, Ľ. Hujo, Journal of Central Euro. Agric., 14, 4 (2013)
15. Z. Tkáč, Ľ. Hujo, J. Tulík, J. Kosiba, D. Uhrinová, V. Šinský, Acta Universitatis Agriculturae et Silviculturae Mendelianae Brunensis, **62**, 5 (2014)
16. Š. Čorňák, In 17th Intern. Scie. Conference Engi. for Rural Develo., Jelgava, Latvia, 2048–2053 (2018)
17. Z. Tkáč, Š. Čorňák, V. Cviklovič, J. Kosiba, J. Glos, J. Jablonický, R. Bernát, ATA, **20**, 2 (2017)

# Evaluation of linear losses in ventilation ducts with a rounded rectangle cross-section

*Kazimierz* Peszynski[1,*], *Václav* Tesař[2] and *Ewa* Kuliś[1]

[1]UTP University of Science and Technology, Al. Prof. S. Kaliskiego 7, Bydgoszcz, Poland
[2]Václav Tesař, Institute of Thermomechanics of the CAS, v. v. i., Dolejškova 1402/5, 182 00 Praha 8, Czech Republic

**Abstract.** The paper presents a new ventilation system characterized by the cross-section of the channels in the shape of a rounded rectangle. The basic research problem was the determination of local losses in the system components. To achieve this goal, it is necessary to know the linear losses in the ducts. The principle of determining both types of losses was presented. The friction coefficients of linear pressure losses referred to the unit of the straight length of the duct section $\zeta_l \left[ \mathrm{m}^{-1} \right]$ as well as generalized dimensionless friction coefficient of linear pressure losses $\lambda \left[ - \right]$ is presented. The main achievement of the paper is the statement that the lambda coefficient $\lambda$ can replace the coefficients $\zeta_l$ for all 79 examined duct cross-sections.

## 1 Introduction

In ventilation systems, as in air conditioning, the basic sections are rectangular and circular sections. Both types have their advantages and disadvantages. In this work, rectangular ducts with rounded corners are analysed, this cross-section is to combine the advantages of both of the above-mentioned shapes of ventilation ducts.

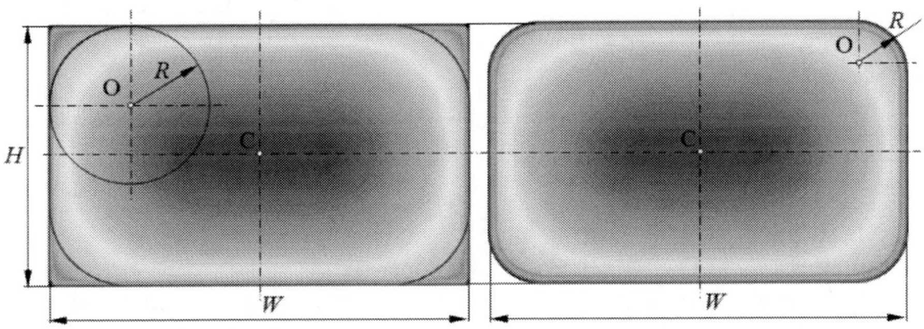


**Fig. 1.** Genesis of the cross-section: a) initial rectangular b) final rounded rectangular corners.

Figure 1 shows the distribution of air velocity in a rectangular section with dimensions $W \times H = 0{,}8\,\mathrm{m} \times 0{,}5\,\mathrm{m}$ obtained during simulation in the ANSYS-FLUENT software code, where the average velocity $v_{avg\_rec} = 5\,\mathrm{m \cdot s^{-1}}$ in the duct cross-section was entered at the entrance to the duct. The maximum velocity was observed at the point of intersection of the axis of symmetry C, it was $v_{max} = v_{Crec} = 6{,}186\,\mathrm{m \cdot s^{-1}}$. For the same rectangle, but with corners rounded with a radius $R = 0{,}1\,\mathrm{m}$, was obtained $v_{max} = v_{Crr} = 6{,}154\,\mathrm{m \cdot s^{-1}}$. This is

* Corresponding author: kazimierz.peszynski@utp.edu.pl

not a surprising result, because it was expected at the same average velocity This result obtained by numerical simulation indicates more even section filling at rounded corners.

## 2 Basic parameters of the examined channels

In Figure 2 selected characteristic cross-sections of the studied channels are presented. The dimensions of the channels are given in meters, e.g. 2.0 × 1.2 means 2.0 m × 1.2 m. The coefficient before $A_{rec} = W \cdot H$ means the indicator of the lost surface of the rectangle as a result of rounding its corners. This coefficient was determined from the formula $A_{rr} / A_{rec}$ .

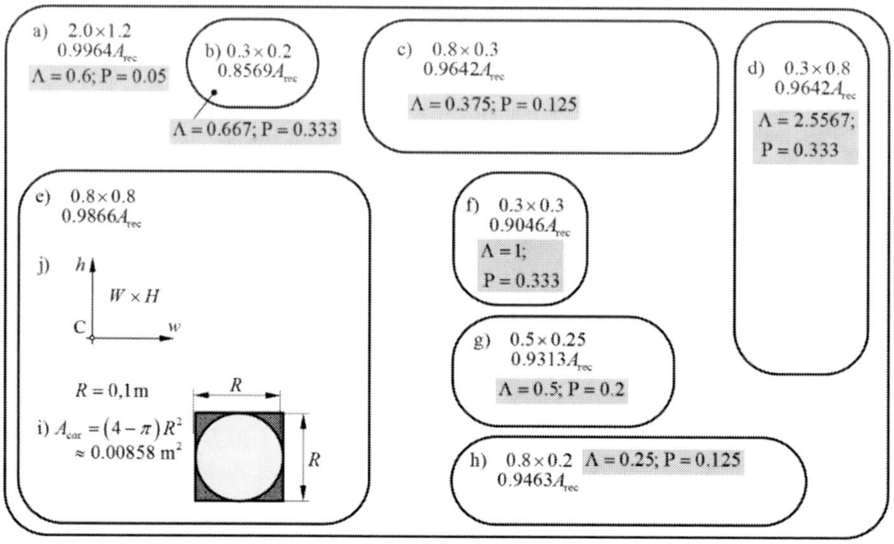


**Fig 2.** Selected ventilation ducts: a) maximum cross-section, b) minimum cross-section, c), d) maxi–mum cross-section adjustable horizontally and vertically, e) maximum square cross-section, f) mini–mum square cross-section, g) cross-section for preliminary tests, h ) the most flattened maximum cross-section, i) determination of the lost area rectangle as a result of rounding the cross-section, j) coordinate system.

Since, for technological reasons, the corner rounding radius was the same and was $R = 0.1\,\text{m}$ (Fig. 2i), a constant value $A_{cor}$ was subtracted from the area $A_{rec} = W \cdot H$ , therefore $A_{rr} = A_{rec} - A_{cor}$ .

The basic geometrical data of the tested duct cross-section and relative parameters are shown in Figure 3. The relative introduced parameters Λ and P allow the generalization of test results, in accordance with the general trend of entering dimensionless numbers.

The cross-sectional area of the rounded rectangle was determined from the formula:

$$A_{rr} = W \cdot H - (4-\pi) R^2 \quad \text{m}^2 \tag{1}$$

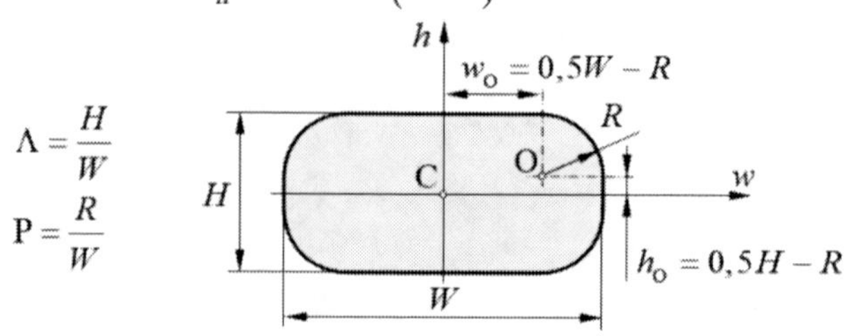


**Fig. 3.** Basic geometrical parameters of the examined cross-section.

The cross-sectional area can also be represented in a parametric form $A_{\text{пr}} = f(\Lambda, \mathrm{P}, W)$, thus determination of area $A_{\text{пr}}$ is important when generalizing the results. The parametric form of the surface area was obtained after substituting the parameters specified in the Figure 3 into the equation (1):

$$A_{\text{пr}} = \left(\Lambda - (4-\pi)\mathrm{P}^2\right)W^2 \quad \mathrm{m}^2 \tag{2}$$

The expression in outer round brackets in the equation (2) is dimensionless and can be treated as a parameter $k_{\mathrm{A}} = \Lambda - (4-\pi)\mathrm{P}^2$ characterizing the channel cross-section depending on the channel width $W^2$. The factor $k_{\mathrm{A}}$ numerically determines how much the cross-section of the examined duct is smaller than the cross-section of the square with side $W$.

Table 1 presents a list of all cross-sections of the ducts new system under examination. The numerical values in Table 1 were determined on the basis of formula (1), for $R = 0{,}1\ \mathrm{m}$.

**Table 1.** List of 79 cross-section area of the examined new system.

| $A_{\text{пr}}$ [m²] | | W [m] | | | | | | | | | | |
|---|---|---|---|---|---|---|---|---|---|---|---|---|
| | | 0,30 | 0,40 | 0,50 | 0,60 | 0,80 | 1,00 | 1,20 | 1,40 | 1,60 | 1,80 | 2,00 |
| H [m] | 0,20 | 0,051 | 0,071 | 0,091 | 0,111 | 0,151 | | | | | | |
| | 0,25 | 0,066 | 0,091 | 0,116 | 0,141 | 0,191 | 0,241 | | | | | |
| | 0,30 | 0,081 | 0,111 | 0,141 | 0,171 | 0,231 | 0,291 | 0,351 | | | | |
| | 0,40 | 0,111 | 0,151 | 0,191 | 0,231 | 0,311 | 0,391 | 0,471 | 0,551 | | | |
| | 0,50 | 0,141 | 0,191 | 0,241 | 0,291 | 0,391 | 0,491 | 0,591 | 0,691 | 0,791 | | |
| | 0,60 | 0,171 | 0,231 | 0,291 | 0,351 | 0,471 | 0,591 | 0,711 | 0,831 | 0,951 | 1,071 | 1,191 |
| | 0,80 | 0,231 | 0,311 | 0,391 | 0,471 | 0,631 | 0,791 | 0,951 | 1,111 | 1,271 | 1,431 | 1,591 |
| | 1,00 | 0,291 | 0,391 | 0,491 | 0,591 | 0,791 | 0,991 | 1,191 | 1,391 | 1,591 | 1,791 | 1,991 |
| | 1,20 | 0,351 | 0,471 | 0,591 | 0,711 | 0,951 | 1,191 | 1,431 | 1,671 | 1,911 | 2,151 | 2,391 |

Hydraulic diameter $D_{\mathrm{H}}$ of rounded rectangular (Fig. 3) was determined from equation:

$$D_{\mathrm{H}} = \frac{4A_{\text{пr}}}{U_{\text{пr}}} = \frac{4\left(H \cdot W - 4\left(1 - \frac{\pi}{4}\right)R^2\right)}{2\left(W + H - (4-\pi)R\right)} = \frac{2\left(\Lambda - (4-\pi)\mathrm{P}^2\right)}{1 + \Lambda - (4-\pi)\mathrm{P}} W \tag{3}$$

where: $A_{\text{пr}}$ is area of rounded rectangular, $U_{\text{пr}}$ – perimeter of rounded rectangular, $\Lambda$ – dimensionless height, $\mathrm{P}$ – dimensionless radius, $W, H, R$ – geometrical parameters of the rounded rectangle.

The right side of the equation (3) is interesting. The fraction before the variable $W$ contains only dimensionless construction parameters: the degree of flattening of the cross-section $\Lambda$ and the degree of rounding of the corners $\mathrm{P}$. This is a number factor which, when multiplied by the section width, gives the hydraulic diameter for this cross-section.

## 3 Linear losses

The main task during the research was to determine local losses for individual system elements. The idea of determining local losses is presented in Figure 4.

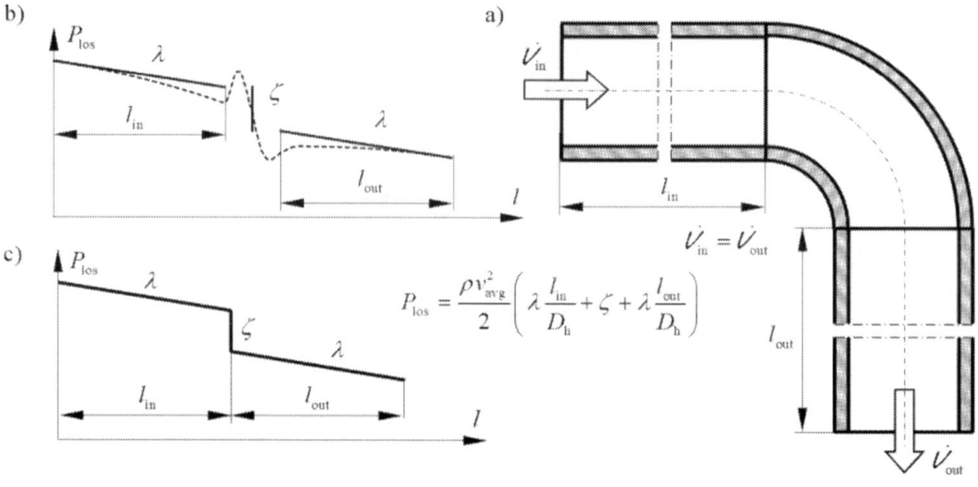


**Fig. 4.** Basic geometrical parameters of the examined cross-section.

Figure 4 shows a ventilation duct with an arc, at which local losses should be determined $\zeta$. A direct measurement of these losses is not possible. It is necessary to prepare a measuring system containing: an input channel of length $l_{in}$, the measured element characterized by the sought local losses $\zeta$, and an output channel of length $l_{out}$. In the literature it is assumed that $l_{in} > 10D_h$ and $l_{out} > 10D_h$. The pressure losses in the system are given by the equation

$$P_{los} = \frac{\rho v_{avg}^2}{2}\left(\lambda \frac{l_{in}}{D_h} + \zeta + \lambda \frac{l_{out}}{D_h}\right) \tag{4}$$

It is assumed that the component of the ventilation system causes sudden losses, i.e. it has zero length – compare the figures Fig. 4b) and Fig. 4c). The average velocity $v_{avg}$ value in the channel is also necessary, its determination was the basic problem during the tests due to the wide range of measurements. Note that the expression $0{,}5\rho v_{avg}^2$ is kinematic pressure. During the research, the effect of air temperature and atmospheric pressure on pressure loss was also analysed. This impact was found to be insignificant. The impact of the average velocity $v_{avg}$ measurement inaccuracy was much larger. In view of the above, this effect is omitted further in the study. From the formula (5) it is possible to determine the local loss coefficient in the examined component

$$\zeta = \frac{2P_{los}}{\rho v_{avg}^2} - \frac{\lambda}{D_h}\left(l_{in} + l_{out}\right) \tag{5}$$

From formula (6) it follows that in order to determine the local loss coefficient $\zeta$, it is necessary to know the linear loss coefficient $\lambda$.

Initially, attempts were made to determine linear losses theoretically. There are different theoretical ways to determine this coefficient $\lambda$, the most popular is the phenomenological Colebrook–White equation, obviously used for turbulent flow in smooth and rough pipes.

$$P_{los} = \frac{\rho v_{avg}^2}{2}\left(\lambda \frac{l_{in}}{D_h} + \zeta + \lambda \frac{l_{out}}{D_h}\right) \tag{6}$$

where: $k$ is roughness of duct, $D_H$ – hydraulic diameter, Re – Reynolds number.

Equation (6) has an implicit nature, the result can be obtained only by iterative methods, which in the current state of computational technology do not cause any problems. However, when calculating using this formula, it seems that it applies to hydraulically smooth channels only. This, however, cannot be said about the ducts of the system under investigation. These ducts consist of segments of length $l = 1.5\,\mathrm{m}$ or $l = 1\,\mathrm{m}$. There is a seal between the segments, which, according to observations resulting from the conducted tests, has a much greater influence on the linear losses coefficient $\lambda$ than that occurring in (1), the surface roughness $k$ of the plates from which the ducts were made. In addition, the sealing effect is stochastic. Therefore, during the tests, $\lambda$ was determined experimentally. The principle of this measurement is illustrated in the Figure 5.

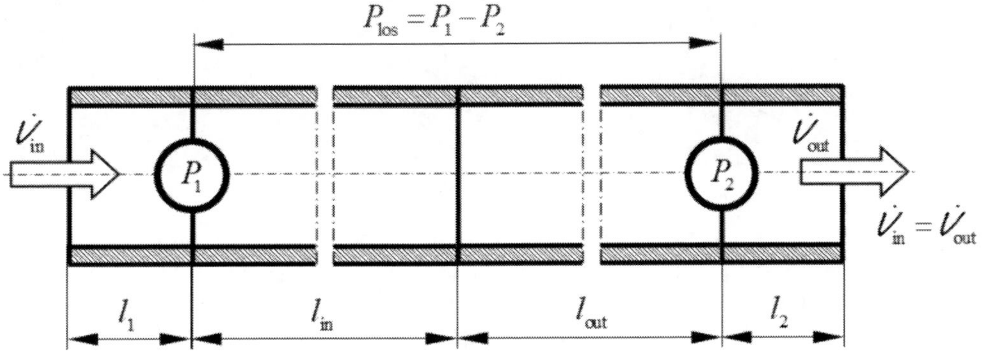


**Fig. 5.** The idea of measuring the linear loss coefficient $\lambda$.

After removing the tested component (in the situation shown in Fig. 4 segment arc), the inlet and outlet parts are combined into stright one with length $l_{in} + l_{out}$. After omitting $\zeta$ in the formula (4) and after its transformation we get the coefficient of local losses

$$\lambda = \frac{2P_{los}D_H}{\rho v_{avg}^2 \left(l_{in} + l_{out}\right)} \tag{7}$$

where: $l_{in} + l_{out}$ is the length of the straight duct, $P_{los}$ – pressure drop along the length of the duct, $D_H$ – hydraulic diameter, $\rho$ – air density, $v_{avg}$ – air average velocity in duct.

The equation (6) illustrates the importance of determining the average flow velocity in the ventilation system duct, therefore the determination of this value publications was devoted to particular attention.

Figure 6 shows the test stand when determining the local loss coefficient for a 60° segmented elbow for ducts with dimensions $W = 0{,}6\,\mathrm{m}$, $H = 0{,}2\,\mathrm{m}$. In the upper left corner (Fig. 6a) you can see the straightener of the supply system and the pressure measuring point at the channel inlet. The hydraulic diameter of this channel is $D_h = 0{,}312\,\mathrm{m}$, i.e. the length $l_{in} > 3{,}12\,\mathrm{m}$, in fact $l_{in} = 6{,}1\,\mathrm{m}$ - 4 segments with a length of 1.5 m, and 0.1 m on the fifth segment. The output channel had the same length, only the fragment with the computer standing on it is visible in the figure. The Fig. 6b) shows a power supply unit whose basic elements are a fan and a spotlight with a chamber.

**Fig. 6.** Test stand before 60° segmented elbow removal.

Figure 6c) shows the pressure measuring point on the axis of the ventilation duct, the basic element is a Prandtl tube. The tube was carefully positioned parallel to the axis of the duct. Fig. 6d) shows the cross-section of the straightener. The straightener channels had the shape of a regular hexagon with a side $a = 5\,\mathrm{mm}$, sheet thickness 0.8 mm.

Table 1 shows linear losses marked with the symbol $\zeta_l$, for all 79 experimental cross-sections tested, given by the formula

$$\zeta_l = \frac{\lambda}{D_\mathrm{h}} = \frac{2P_\mathrm{los}}{\rho v_\mathrm{avg}^2 \left(l_\mathrm{in} + l_\mathrm{out}\right)} \quad \left[\mathrm{m}^{-1}\right] \tag{8}$$

**Table 2.** Values of measured linear loss coefficients $\zeta_l$ for 79 cross-section area.

| $\zeta_1\left[\mathrm{m}^{-1}\right]$ | | W [m] | | | | | | | | | | |
|---|---|---|---|---|---|---|---|---|---|---|---|---|
| | | 0.3 | 0.4 | 0.5 | 0.6 | 0.8 | 1.0 | 1.2 | 1.4 | 1.6 | 1.8 | 2.0 |
| H [m] | 0.20 | 0.099 | 0.085 | 0.077 | 0.072 | 0.068 | | | | | | |
| | 0.25 | 0.092 | 0.078 | 0.070 | 0.063 | 0.056 | 0.052 | | | | | |
| | 0.30 | 0.084 | 0.073 | 0.066 | 0.060 | 0.052 | 0.048 | 0.045 | | | | |
| | 0.40 | 0.073 | 0.064 | 0.058 | 0.053 | 0.046 | 0.041 | 0.038 | 0.036 | | | |
| | 0.50 | 0.062 | 0.056 | 0.050 | 0.046 | 0.040 | 0.035 | 0.032 | 0.030 | 0.029 | | |
| | 0.60 | 0.054 | 0.050 | 0.045 | 0.041 | 0.036 | 0.031 | 0.028 | 0.026 | 0.024 | 0.023 | 0.022 |
| | 0.80 | 0.049 | 0.045 | 0.041 | 0.037 | 0.031 | 0.026 | 0.023 | 0.021 | 0.020 | 0.019 | 0.018 |
| | 1.00 | 0.045 | 0.042 | 0.038 | 0.034 | 0.028 | 0.023 | 0.020 | 0.018 | 0.017 | 0.016 | 0.016 |
| | 1.20 | 0.042 | 0.039 | 0.036 | 0.032 | 0.026 | 0.022 | 0.019 | 0.017 | 0.016 | 0.015 | 0.015 |

This form of linear losses coefficient presentation is very popular in the catalogs of ventilation system published by manufacturers. Graphic form of $\zeta_l(w)_{h=const}$ for selected $h$ is presented also in Fig 8 – dashed lines. Losses in the duct are determined by multiplying the value $\zeta_l$ by the length of the duct. Each duct cross-section area has a different linear loss value $\zeta_l$. These are the direct results of the experimental studies described above.

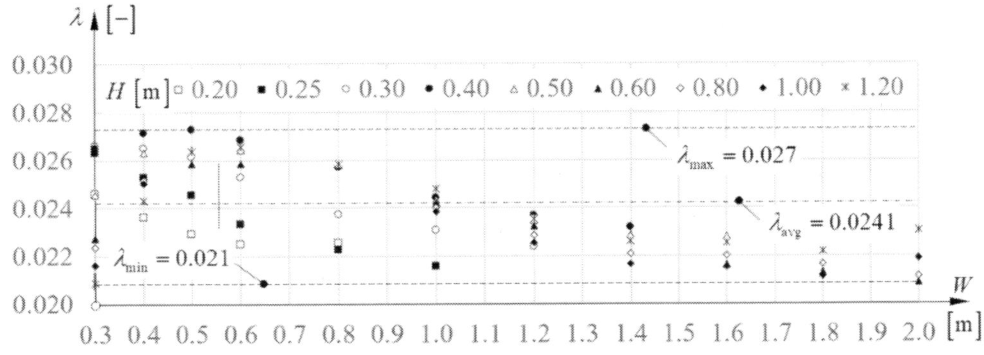


**Fig. 7** Dimensionless pressure loss coefficient $\lambda$ .

Dimensionless linear loss coefficients $\lambda$ – equation (6) – has a lot more general meaning. In order to obtain a dimensionless pressure loss coefficient $\lambda$ based on the data presented in Table 1, it is necessary to multiply this data by the appropriate hydraulic diameter $D_H$ determined by equation (4). Values of $W$ and $H$ were taken based on the coordinates of the table 2, the parameter $R = 0.1$ was constant for all cross-sections. The result of this multiplication is shown in Fig. 3 in the form of a graph.

The values shown in Figure 7 are within a narrow range of values $\lambda \in (0.020;\ 0.027)$. If we accept as true the sentence included in one of the fundamental textbook on fluid mechanics (Cengel et all, 2006), chap. 8, p. 322 – "*An error of 10 percent (or more) in friction factors calculated using the relations in this chapter is the "norm" rather than the "exception.*" – it is reasonable to calculate the average value for all 79 duct shapes. Therefore

$$\lambda_{avg} = \frac{1}{79}\sum_{i=1}^{79}\lambda_i = 0{,}024 \tag{9}$$

The value of dimensionless linear loss coefficient $\lambda = \lambda_{avg}$ determined in the above manner was then used for any comparative analysis of ventilation ducts, first of all for determining local losses coefficient $\zeta$ of system fittings.

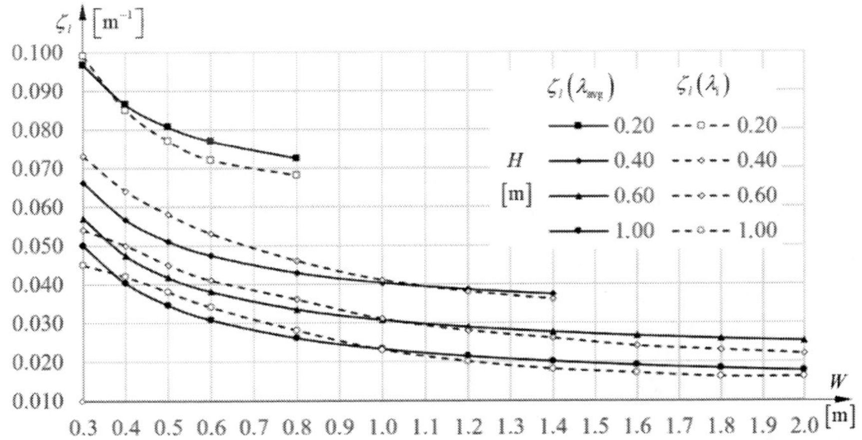


**Fig. 8** Comparison of linear loss coefficients $\zeta_l$ measured directly and determined on the basis of averaged dimensionless coefficient $\lambda$ .

Figure 8 presents a comparison of $\zeta_l$ coefficients measured directly during tests with $\zeta_l\left(\lambda_{avg}, D_H\right)$ coefficients calculated on the basis of average $\lambda_{avg}$ lambda coefficient and hydraulic diameter $D_H$. For clarity of the image, only 4 sets of results from the available 9 are shown. The calculated coefficients are marked with a solid line and black markers, while the values measured ones with a dashed line and with corresponding white markers. The curves presented in this figure confirm the thesis that all 79 $\zeta_l$ coefficients can be replaced by one $\lambda$ coefficient.

The reader based on Table 2 and $\lambda = 0.024$ can at any time make the missing 5 sets of curves using the left side of the equation (7). This conclusion was very useful during later experimental testing of local pressure losses.

## Conclusions

The dimensionless linear loss coefficient $\lambda[-]$ can replace all dimensional loss coefficients $\zeta_l\left[\mathrm{m}^{-1}\right]$ for 79 examined rounded rectangle duct cross-sections.

The advantage of the above procedure is assimilation experimentally obtained dimensionless linear loss coefficient $\lambda$ to the coefficient determined from the Colebrook White formula (1). In this equation, only the Reynolds number $Re$, the roughness of the plate $k$ and the hydraulic diameter $D_H$ of the cross-section are present, but there is no cross-sectional shape.

The Colebrook–White equation can be applied to a duct with a different cross-section than circular, one should remember about the influence of seals between the duct segments. This goal can be achieved by the correction of the roughness coefficient $k$.

The authors would like to kindly thank the company Nucair Technologies Sp. z o.o., Solec Kujawski, Poland for the performance of tested ducts and delivery of the air supply system. Authors also received institutional support BS 28/2018 granted by Faculty of Mechanical Engineering of UTP University.

## References

1. Y.A, Cengel, J.M. Cimbala: Fluid Mechanics. Fundamentals and Applications, McGraw-Hill, (2006)
2. K. Peszyński (2018) *Building a bridge between industry and theory on the example of a new ventilation system*. EPJ Web of Conferences Volume 213 (2019)
3. K. Peszyński, J. Novosád, E. Smyk, Ł. Olszewski, P. Dančová (2017) *Modelling of air flow rate in significantly flattened rounded rectangular ventilation ducts*., EPJ Web of Conferences 180, 02082 (2018)
4. K. Peszyński, L. Olszewski, E. Smyk, and D Perczyński. (2016) *Analysis of the Velocity Distribution in Different Types of Ventilation System Ducts*, EPJ Web of Conferences 180, 02081 (2018)
5. K. Peszynski, L. Olszewski, E. Smyk, and T. Kasprowicz (2017): Development of New Type Ventilation Ducts System, Proceedings of 23rd International Conference Engineering Mechanics 2017, p. 50 – 53, ISBN 978-80-214-5497-2, ISSN 1805-8248.
6. K. Peszyński, D. Perczyński, E. Smyk, P. Kolber (2017): *Experimental Verification of Velocity Distribution in Different Cross-sectional Ventilation Ducts*, Proceedings of 23rd International Conference Engineering Mechanics 2017, p. 770 – 783 2017, ISBN 978-80-214-5497-2, ISSN 1805-8248, (2017)

7. E. Smyk, D. Mrozik, Ł. Olszewski, K. Peszyński (2017) *Numerical simulation of minor losses coefficient on the example of elbows*, EPJ Web of Conferences 180, 02093 (2018)
8. E. Smyk, P. Mrozik, S. Wawrzyniak, K. Peszyński (2017): *Tabular Air Deflector in Ventilation Ducts*, Proceedings of 23rd International Conference Engineering Mechanics 2017, p. 882 – 885, ISBN 978-80-214-5497-2, ISSN 1805-8248, (2017)

# Diagnostics of emission condition of compression ignition engine in utilization of tested substance

*Ján* Polerecký[1], *Juraj* Jablonický[2,*], *Juraj* Tulík[2], *Ján* Kosiba[2], *Marcin* Zastempowski[3], and *Mirko* Simikić[4]

[1]S-EKA spol. s r.o., authorized organization for emission control of road vehicles, Kupecká 5, 949 01 Nitra, Slovak Republic
[2] Faculty of Engineering, Slovak University of Agriculture in Nitra, Tr. A. Hlinku 2, 949 76 Nitra, Slovak Republic
[3] Faculty of Mechanical Engineering, UTP University of Science and Technology in Bydgoszcz, al. Prof. S. Kaliskiego 7, 85-796 Bydgoszcz, Poland
[4] Faculty of Agriculture, University of Novi Sad, Trg. Dositeja Obradovića, 21101 Novi Sad, Serbia

**Abstract.** This paper focuses on the observation of a decrease in smoke opacity values in application of tested substance in selected vehicles with a compression ignition engine. The aim of the paper is to evaluate the ability of the tested substance to decrease the smoke opacity value in vehicles equipped with the compression ignition engine with mileage above 30,000 km. Measurements were carried out on the basis of determined measurement methods. From the results obtained in measurements on two vehicles, it can be concluded that the utilization of the tested substance led to a decrease in the smoke opacity value in both observed vehicles. The results of the experiment show further possibility for decreasing of emission production in exhaust gases of motor vehicles equipped with the compression ignition engine.

## 1 Introduction

Developments in the field of technology bring with their positive impact and increase of life quality also undesirable side effects. One of the severe adverse effects of the scientific and technical progress is environmental pollution [1, 2, 3, 4, 5, 6]. The transport sector has a significant influence on worsening ecological situation as it not only particulates in ecosystem pollution but is also one of the main causes for the existence of global energy issues. We can now witness the producers' efforts to increase the ecological safety taking place as early as in the motor vehicle development. As a result, vehicles are equipped with exhaust systems supplemented with exhaust gas treatment equipment. Engine developers and producers can also utilize alternative fuels (CNG, LPG, E85, biodiesel) as an energy source for vehicles, as well as transition to hybrid propulsion (combustion and electric motor) and introduction of fuel cells. Another option for increasing of ecological safety is an application of electric drives. In evaluating the ecological impact of vehicle operation, it is necessary to take into account all the factors participating in environmental load, from production of the vehicle, its drive units and accessories up to disposal of the vehicle.

With regard to the need for limiting air pollution by combustion engine exhaust gases, the maximum production of pollutants in exhaust gas emissions of all motor vehicles is legally limited. The EU legislature sets the maximum permissible values of exhaust gas emissions (EURO limits) during the type tests for general authorization to operate (homologation) and during the pollutant emission checks (emission checks) of vehicles in

*Corresponding author: juraj.jablonicky@uniag.sk

operation. Emission limits are expressed in terms of engine performance and validity period, given in $g.kW^{-1}.h^{-1}$. Limited are CO, hydrocarbons, volatile organic HC compounds, suspended PM, and $NO_x$ oxides. With growing strictness of the limits, the sulfur content in diesel fuel has been decreasing over the years. The lowering of sulfur content is relevant from the standpoint of the technologies utilized for further emission treatment, such as solid particulate filters, in which sulfur reduces the efficiency of performance and lifespan [7]. Engine emissions contain hundreds of chemicals with different concentrations, the biological properties (impact on human health) of which have not been exactly determined to date. In case additives are used in fuel, applied directly to the working area of the combustion engine, new types of substances, usually not present in exhaust gases, can be generated, further contributing to the catalytic chemical reactions. These are for example some of the fuel additives – "fuel-borne catalysts", used to support the regeneration of particulate filters and connected to the production of highly toxic dioxin and furan emissions [8]. The possibility of new substance generation needs to be taken into account also when additives (e.g. catalytic) are applied to fuel or lubricating oils and when liquids are added to exhaust gases. A common example is urea used for reduction of $NO_x$ emissions in SCR (Selective Catalytic Reduction) catalysts – in such a case SCR may contain ammoniac and multiple other substances from incomplete decomposition of urea. Combustion engines are responsible for more than 70% of global CO emissions production and 19% of global $CO_2$ emissions production. In addition to the products of perfect combustion – i.e. $CO_2$, $H_2O$, excess oxygen, excess nitrogen – which represent the majority of exhaust products, a wide range of other gases and solids may occur, which tend to hold greater attention: CO, non-combusted hydrocarbons (paraffins, olefins, aromatic hydrocarbons), partially combusted hydrocarbon (aldehydes and ketones), degradation products (acetylene, ethylene, hydrogen, soot), nitrogen oxides $NO_x$ (NO – nitrogen monoxide, $N_2O$ – dinitrogen monoxide, $NO_2$ – nitrogen dioxide) and solid particulate matter. The degree of harmfulness of individual exhaustion gas elements is sometimes given by means of comparison to the harmfulness of CO. Providing objective figures for levels of harmfulness is certainly difficult. It is evident from the operation of a compression ignition engine that the increase in smoke opacity leads to an increase in production of pollutants (CO and HC), the measuring of which is difficult in practice, however. It is therefore vital for practice to observe the value of particulate emissions PM (Particulate Matter), the measuring of which is much faster and technologically and economically simple, while simultaneously being sufficient for evaluation of technical condition. Compression ignition engines operate narrowly below the smoke opacity threshold during the maximum performance [9]. Engine operation results in production of smoke, generated due to insufficient oxygenation of the fuel in the combustion cycle, i.e. due to insufficient combustion. Smoke production is thus caused either by lack of airflow or excess of fuel. Both causes are invariably bound with the technical condition of the compression ignition engine.

One of the ways to comply with the stricter emission regulations is to focus on and search for suitable alternative fuels. The main plausible alternative fuels used in car transport are ethanol, hydrogen and biodiesel [10]. Hydrogen-based fuel cells could become a real alternative to fossil fuels, but for this to be commercially viable, many technical issues need to be solved beforehand, such as intricacy of hydrogen production, requirements for the infrastructure for its storage and high production costs of the fuel cell. A high number of studies have shown that biodiesel could serve as an alternative for compression ignition engines, with small or even no requirements for their adjusting [11]. It was also proven that biodiesel has a great potential for decreasing the $CO_2$, CO, THC (Total Hydro Carbon) emissions and PM emissions Regulation no. 24 [12, 13]. In conclusion, alternative fuels and their mixtures with diesel fuel are still a subject of research, focusing primarily on reduction of emission production arising from their combustion in engine, but also on the transformation of their heat into mechanical energy.

## 2 Measurement of solid particles emission

The method of measurement of solid particle emissions by opacity observation is based on the principle of measuring the physical property of a gaseous environment characterizing the optical absorbability of a gas column of a defined length – i.e. opacity, which is expressed by the value of attenuation coefficient. Opacity, measured by means of an opacimeter, is therefore an inverted value of transparency. The value of opacity (smoke opacity) is expressed by light attenuation coefficient $k$, the determining of which is laid down by UN/ECE Regulation [12] in accordance with UN/ECE Regulation [14, 15] and is calculated by means of Eq. (1):

$$\phi = \phi_O \, . \, e^{-kL} \tag{1}$$

where: $L$ – effective length of the light path in the observed gas, m; $\Phi_O$ – light flux emitted from the observed gas, -; $\Phi$ – incident light flux, -; $k$ – light attenuation coefficient, $m^{-1}$; $e$ – base of natural logarithm.

It can be observed from Eq. 1 that the attenuation coefficient $k$ (Eq. 2) is a logarithmic function of the ratio of lur

$$k = -\frac{1}{L}\frac{\log\frac{\Phi}{\Phi_0}}{\log e}, m^{-1} \tag{2}$$

Opacity measurement by the method of free acceleration is defined by Law no. 106/ 2018 for the measurement of particulate emissions used to control the emission condition of the diesel engine, as well as for homologation tests of road motor vehicles [16]. Measurement conditions are laid down by Annex IV of Council-Directive 2014/45/EU.

### 2.1 Equipment used in measurements

In measurements of the emission condition of vehicles while utilizing the tested substance, the following measuring devices were used:

1. Speed measurement device DiSpeed 492 by AVL, registered at Graz, which is intended to be used for measuring of engine speed of both two- and four-stroke engines. Speed scanning is based on measurement of signal caused by engine vibrations or of sound vibrations scanned by a sensor applied to the engine of the controlled vehicle by means of a permanent magnet.
2. MAHA MGT 5 and MDO – 2 Lon Euro emission tester by MAHA Maschinenbau Haldenwang GmbH & Co. KG Germany, which meets the requirements set by UN/ECE Regulation no. 24 and ISO 11614:1999 [12, 17](Reciprocating internal combustion compression-ignition engines – Apparatus for measurement of the opacity and for determination of the light absorption coefficient of exhaust gas).

#### *2.1.1 Measurement method*

1. Carry out first measurements for obtaining of input values of smoke opacity and for evaluation of emission condition of the vehicles without the tested substance.
2. Repeat measurement of vehicle emission condition without the tested substance after 1,000 km (permissible deviation +200 km).
3. Determine the minimal required number of measurements by means of mathematical statistical analysis from the obtained input values.
4. Determine the dosage ratio.
5. Carry out the vehicle emission condition measurements with the tested additive substance – after 500 km (permissible deviation +100 km).
6. Carry out every subsequent measurement with the tested substance after additional 600 km (permissible deviation ±100 km).

7. Evaluation of the results obtained.

### 2.1.2 Mathematical statistical analysis

Determining the necessary number of experiment repetitions can be, according to the listed methods, carried out when the standard deviation of the observed property or variation coefficient are known. These values can be known from previous measurements, from preliminary control measurements or literature. If the standard deviation is known, it is possible to determine the necessary number of experiment repetitions for selection with repetition on the basis of Eq. 3:

$$n = t_a^2 . \frac{s}{\Delta^2} \quad (3)$$

where: $n$ – the number of repetitions being determined; $t_a$ – the critical value of normal distribution for bilateral estimate; Δ – the selected value of the required accuracy;
$s$ – the estimate of the standard deviation of the core set.
In determining the necessary number of repetitions of the selection set, Eq. 4 provided by [18] can be used:

$$n = \frac{N.t_a^2.s^2}{t_a^2.s^2 + N.\Delta} \quad (4)$$

where: $N$ – the size of the set used.

If the variation coefficient of the set $V_k$ is known, then the necessary number of repetitions can be determined by

$$n = \frac{v_k^2.t_a^2}{\delta^2} \quad (5)$$

where: $n$ – the observed number of repetitions; $t_a$ – critical value of normal distribution for bilateral estimate; $v_k$ – variation coefficient, %; $\delta$ – maximum permissible error, %.

## 2.2 Characteristic of the tested substance

The tested substance is designed for additivation of diesel fuel of all types according to EN 590. The ferrocene active substance influences the combustion process and in perfect combustion lowers the exhaust gas emissions. The tested substance decarbonizes the combustion engine and maintains the cleanliness of the fuel system. It decarbonizes the area of the first piston ring and nozzle outlet. After decarbonization, the combustion chamber becomes covered by a brightly red microscopic protective layer and does not produce any harmful compounds.

### 2.2.1 Criteria for evaluation of influence of the tested substance

The following criteria were set for evaluation of the influence of the tested substance on the emission parameters of the vehicles:
1. The mileage of the test vehicles must be at least 30,000 km.
2. The vehicles must be fully operational in road traffic.

Table 1 contains the basic characteristics of the observed vehicles to which the substance for emission lowering was applied.

**Table 1.** Characteristics of the tested vehicles.

| Mark | Vehicle 1 | Vehicle 2 |
|---|---|---|
| Cylinder capacity | 2,287.2 $cm^3$ | 2,999.0 $cm^3$ |
| Catalytic converter | KAT | KAT |
| Maximum engine power/engine speed | 88.00 kW/3,600 $min^{-1}$ | 115.50 kW/3,500 $min^{-1}$ |
| Fuel type/energy source | diesel | diesel |

| | | |
|---|---|---|
| Transmission/number of gears | MT6 | MT6 |
| Operating weight | 2,060 kg | 2,215 kg |
| Maximum permissible total weight | 3,500 kg | 3,500 kg |
| **Emissions and consumption** | | |
| Emissions EC/UNECE | 70/220*2003/76B (EURO 4) | 70/220*2003/76B (EURO 4) |
| | | |
| SMOKE OPACITY EC/UNECE | 72/306*2005/21 | 72/306*2005/21 |
| CO | 0.091 g.km$^{-1}$ | 0.069 g.km$^{-1}$ |
| $NO_x$ | 0.352 g.km$^{-1}$ | 0.350 g.km$^{-1}$ |
| $HC + NO_x$ | 0.396 g.km$^{-1}$ | 0.362 g.km$^{-1}$ |
| Particles | 0.053 g.km$^{-1}$ | 0.056 g.km$^{-1}$ |
| Corrected attenuation coefficient | 0.970 m$^{-1}$ | 1.540 m$^{-1}$ |
| $CO_2$ | 220 g.km$^{-1}$ | 246 g.km$^{-1}$ |

## 3 Results and discussions

For the objective assessment of the tested substance from the perspective of smoke opacity value reduction, it was necessary to determine the minimum sufficient number of experiment repetitions by means of mathematical and statistical analysis. Input data for the determination of the number of repetitions were obtained from smoke opacity measurements on the sample of vehicles equipped with a compression ignition engine with mileage over 30,000 km.

From the experience gained during the measurements and continuous evaluation of the measured data, the number of test repetitions was determined on the basis of variation coefficient, according to Eq. 5. The calculation is based on the maximum permissible error $\delta = 10\%$ and critical value $t_a$ = 1.64 determined with a probability of 90%, which is considered sufficient in experiments within machine engineering (Regulation no. 49). The calculated value n = 67 represents the minimum number of necessary test repetitions.

Vehicle no. 1, the essential parameters of which are shown in Table 1, has the maximum smoke opacity of 1.47 m$^{-1}$ according to the regulations provided by the manufacturer. It was necessary to maintain the range of idling and maximum engine speed specified by the vehicle's manufacturer during testing. Before the application of tested sample to vehicle's fuel system, initial measurements were performed in order to assess the technical condition of the compression ignition engine of the tested vehicle. As evident from Table 2, the combustion engine in vehicle no. 1 is in adequate technical condition; this is indicated by the mean smoke opacity value determined after two measurement cycles realized after a 1,100 km run. The mean smoke opacity value was 0.46 m$^{-1}$, i.e. 68.9% less than required by legislature.

**Table 2.** Measured input values of smoke opacity for vehicle no. 1.

| **Vehicle no. 1** | | | | | | | | | | | |
|---|---|---|---|---|---|---|---|---|---|---|---|
| Obligatory data for testing | | | Engine temperature, °C | | | | 80 | | | | |
| | | | Idling engine speed, $min^{-1}$ | | | | from 500 to 1,000 | | | | |
| | | | Maximum engine speed, $min^{-1}$ | | | | from 3,500 to 5,500 | | | | |
| | | | Smoke opacity–permissible, $m^{-1}$ | | | | 1.47 | | | | |
| | | | Dispersion, $m^{-1}$ | | | | 0.5 | | | | |
| Number of measurements per one measurement cycle | Number of measurements | Mileage, km | Duration of acceleration, s | Measured temperature, °C | Measured idling engine speed, $min^{-1}$ | Measured maximum engine speed, $min^{-1}$ | Measured smoke opacity, $m^{-1}$ | Mean value of smoke opacity, $m^{-1}$ | Dispersion, $m^{-1}$ | Summary of the mean value of smoke opacity, $m^{-1}$ | Summary of dispersion, $m^{-1}$ |
| ***Measurement without the tested substance*** | | | | | | | | | | | |
| 1 | 1 | 80,048 | 1.55 | 83 | 810 | 4,330 | 0.53 | 0.52 | 0.04 | 0.46 | 0.19 |
| | 2 | | 1.35 | | 800 | 4,370 | 0.49 | | | | |
| | 3 | | 1.42 | | 810 | 4,410 | 0.53 | S | S | | |
| 2 | 4 | 81,148 | 1.41 | 81 | 800 | 4,430 | 0.46 | 0.40 | 0.12 | | |
| | 5 | | 1.48 | | 800 | 4,370 | 0.39 | | | | |
| | 6 | | 1.45 | | 800 | 4,470 | 0.34 | S | S | S | S |

S – satisfactory

After testing the compression ignition engine technical condition in vehicle no. 1, the tested substance was applied to the fuel system in a mixing ratio of 29.5 ml of tested substance per 60 liters of fuel. According to the obtained results, the test itself can be divided into three phases. In the first phase, the mean smoke opacity value before decarbonization was 0.42 $m^{-1}$. The mileage of vehicle no. 1 was 1,110 km at this point of testing. The second phase is represented by the process of decarbonization; it is characterized by an increase in smoke opacity value due to cleaning of the combustion chamber. The mean smoke opacity value was 0.86 $m^{-1}$; the mileage of vehicle no. 1 was 3,010 km in the second phase. The third phase includes the monitoring of the tested vehicle in operation after decarbonization. The mean smoke opacity value in this phase was 0.33 $m^{-1}$ with a mileage of 9,790 km. The graphical dependence of all three phases of the smoke opacity value monitoring with utilization of the tested substance in tested vehicle no. 1 is shown in Fig. 1.

If only the first and third phases were included in the final assessment of the tested substance's impacts on smoke opacity, then the mean smoke opacity value of vehicle no. 1 would be 0.37 $m^{-1}$, representing a decrease in smoke opacity by 74.5% in comparison with the value specified by the manufacturer. When comparing the sums of mean smoke opacity values from the first and third phase with the smoke opacity value measured before application of the tested substance, a 17.9% decrease in the smoke opacity value can be observed.

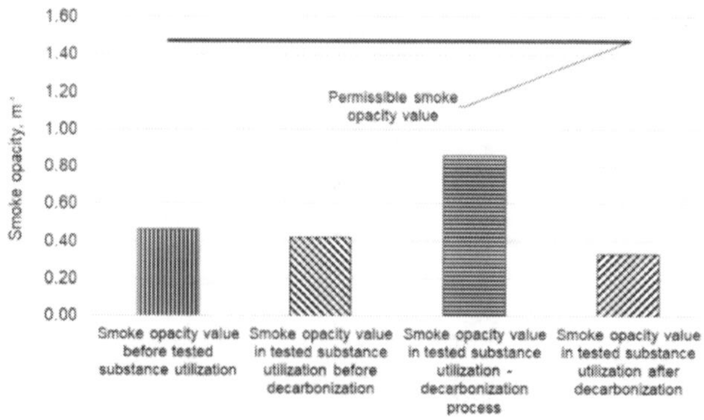


**Fig 1.** Graphical dependence of individual smoke opacity monitoring phases using the test sample in observed vehicle no. 1.

Vehicle no. 2, the essential parameters of which are shown in Table 1, has the maximum smoke opacity of 2.04 $m^{-1}$ according to the regulations provided by the manufacturer. It was necessary to maintain the range of idling and maximum engine speed specified by the vehicle's manufacturer during testing. Before application of the tested sample to vehicle's fuel system, initial measurements were performed in order to assess the technical condition of the compression ignition engine of the tested vehicle. As evident from Table 3, the combustion engine of vehicle no. 2 is in adequate technical condition. This is confirmed by the mean smoke opacity value determined after two measuring cycles realized after a 1,150 km run. The mean smoke opacity value was 2.04 $m^{-1}$, i.e. 49.18% less than required by legislature.

**Table 3.** Measured input values of smoke opacity for vehicle no. 2.

| **Vehicle no. 2** | | | | | | | | | | | |
|---|---|---|---|---|---|---|---|---|---|---|---|
| Obligatory data for testing | | | Engine temperature, °C | | | | 80 | | | | |
| | | | Idling engine speed, $min^{-1}$ | | | | from 500 to 1,000 | | | | |
| | | | Maximum engine speed, $min^{-1}$ | | | | from 3,500 to 4,550 | | | | |
| | | | Smoke opacity – permissible, $m^{-1}$ | | | | 2.04 | | | | |
| | | | Dispersion, $m^{-1}$ | | | | 0.5 | | | | |
| Number of measurements per one measuring cycle | Number of measurements | Mileage, km | Duration of acceleration, s | Measured temperature, °C | Measured idling engine speed, $min^{-1}$ | Measured maximum engine speed, $min^{-1}$ | Measured smoke opacity, $m^{-1}$ | Mean value of smoke opacity, $m^{-1}$ | Dispersion, $m^{-1}$ | Summary mean value of smoke opacity, $m^{-1}$ | Summary value of dispersion, $m^{-1}$ |

| *Measurement without the tested substance* | | | | | | | | | | | |
|---|---|---|---|---|---|---|---|---|---|---|---|
| 1 | 1 | 33,000 | 1.21 | 80 | 810 | 4,100 | 1.09 | 1.08 | 0.04 | 1.04 | 0.11 |
| | 2 | | 1.30 | | 810 | 4,110 | 1.05 | | | | |
| | 3 | | 1.21 | | 810 | 4,110 | 1.09 | S | S | | |
| 2 | 4 | 34,150 | 1.08 | 81 | 800 | 3,604 | 0.98 | 1.00 | 0.03 | | |
| | 5 | | 1.00 | | 799 | 4,200 | 1.00 | | | | |
| | 6 | | 1.02 | | 800 | 4,200 | 1.01 | S | S | S | S |

S – satisfactory

After testing of the compression ignition engine technical condition in vehicle no. 2, the test substance was, similarly to the previous case, applied to the fuel system in a mixing ratio of 29.5 ml of test substance per 60 liters of fuel. According to the results obtained, the test itself can be divided into three phases. In the first phase, the mean smoke opacity value before decarbonization was 0.92 $m^{-1}$. The mileage of vehicle no. 2 was 540 km at this point of testing. The second phase is represented by the process of decarbonization; it is characteristic by an increase in smoke opacity value due to cleaning of the combustion chamber. The mean smoke opacity value was 1.15 $m^{-1}$; the mileage of vehicle no. 2 was 1,840 km in the second phase. The third phase represents the monitoring of vehicle's operation after decarbonization. The mean smoke opacity value in the third phase was 0.95 $m^{-1}$; the mileage of vehicle no. 2 was 10,380 km in this phase. The graphical dependence of all the three phases of the smoke opacity value monitoring with utilization of tested substance in the observed vehicle no. 2 is shown in Fig. 2.

As evident from Fig. 4, the smoke opacity value after application of tested substance before decarbonization decreased by 55.1% in the first phase in comparison with the value specified by the manufacturer. When comparing the values before application of tested substance and before decarbonization in the first phase, the smoke opacity decreased by 11.5%.

There was a decrease in smoke opacity by 43.5% in comparison with the smoke opacity value specified by manufacturer in the second phase (the process of decarbonization). In this phase, the mean smoke opacity value increased by 9.6% in comparison with the mean smoke opacity value before application of tested substance.

In the third phase (decarbonization process), the smoke opacity value decreased by 53.6% in comparison with the smoke opacity value specified by the manufacturer of vehicle no. 2. When comparing the smoke opacity value before application of tested substance with the smoke opacity value after decarbonization in the third phase, smoke opacity decreased by 8.7%.

Similarly, to the previous case, if only the first and third phases were included in the final assessment of the test sample's impacts on smoke opacity, then the mean smoke opacity value of vehicle no. 2 would be 0.93 $m^{-1}$, representing a decrease in smoke opacity by 54.3% in comparison with the value specified by the manufacturer. When comparing the sums of mean smoke opacity values from the first and third phase before the application of tested substance, a 10.6% decrease in smoke opacity value is seen.

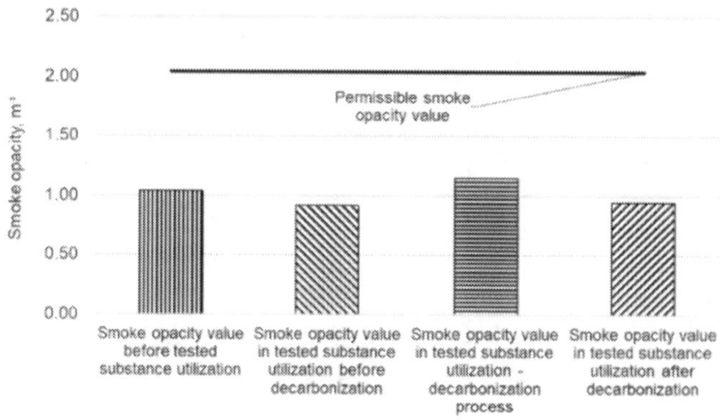


**Fig. 2.** Graphical dependence of individual smoke opacity monitoring phases using the test sample in the observed vehicle.

As already mentioned in the paper's introduction, one of the ways to meet the strict emission regulations is to focus attention on and search for suitable alternative fuels as a source of energy for compression ignition engines. Biodiesel fuel is the most utilized alternative fuel in compression ignition engines. The presented paper aims to point out other options for reduction of particulates in compression ignition engines by application of additives to the fuel system. The application of the tested substance resulted in a reduction of the smoke opacity value in both monitored vehicles by 14.25% on average. However, more significant reduction was observed for biodiesel. The reduction of particulate emissions in the range of 50–70% in the testing of eight types of diesel fuel oil methylester fuels in the fuel system [19]. The next authors [20, 21] also observed that the particulate emission values can be reduced by application of biodiesel to the fuel system. Authors [20], claim that it is possible to reduce particulate emissions by 75–91% by means of biofuel. Papers by [22, 23, 24, 25], who focused their research on monitoring of emissions from compression ignition engines, demonstrate that particulate emission values decreased by more than 50%, in favor of biodiesel over diesel. Authors [26] also confirmed the reduction of particulate emissions by 20% and 50% at 20% and 100% biodiesel utilization. It is evident from the up-to-day results obtained in the compression ignition engine tests with biodiesel that smoke opacity is lower in comparison with the tested substance. However, it is necessary to emphasize that although biodiesel utilization results in reduction of smoke opacity, it also leads to a decrease in vehicle performance. Since neither the tests focused on monitoring of engine performance nor fuel consumption were performed in application of tested substance to the fuel system of compression ignition engine, further measuring and tests must be carried out within this area.

## Conclusions

Nowadays, when our environment is overburdened with emissions of all kinds, every effort leading to reduction of the amount of harmful substances produced by combustion engines is of great significance. Emissions from compression ignition engines have negative impact on both the environment and people. This concerns especially the emissions of pollutants, namely "greenhouse" gases, which cause the gradual irreversible warming of the planet and disturb the balance in nature with acid rains. The major greenhouse gases are considered to be carbon dioxide $CO_2$, oxides of nitrogen $NO_x$, methane $CH_4$ and oxides of sulfur $SO_x$.
The reduction of emissions, defined by regulations, can be achieved by two solutions in designing of motor vehicles. One solution is aimed at adjustment of combustion processes so that the minimum amount of monitored particulates are generated in the combustion chamber; the second solution is aimed at the so-called subsequent emission treatment in the

exhaust pipe after emissions had been generated. The aforementioned solutions belong to groups of engine modifications and engine operating conditions.
Further efforts to reduce the environmental burden lead to the usage of alternative fuels applied in positive ignition and compression ignition engines, hybrid drives and electric generators.

The obtained results presented in this paper demonstrate that the tested substance, which was applied to the vehicle tank in the set mixing ratio resulted in reduction of smoke opacity by 18.0% in vehicle no. 1. The smoke opacity in vehicle no. 2 was reduced by 10.2%. The minimum number of measurements – 67 – was determined on the basis of mathematical and statistical analysis and was respected for both vehicles during the tests. The mileage of vehicle no. 1 was 15,560 km. The mileage of vehicle no. 2 was 14,440 km. In both tested vehicles, an increase in smoke opacity value occurred in a certain phase of monitoring. This is a result of the tested substance impacts – it decarbonizes the combustion chamber and thus maintains fuel system cleanliness. Decarbonization was realized at different intervals in the vehicles, which can be a result of different mileages, as well as different vehicle loads.

This work was supported by AgroBioTech Research Centre built in accordance with the project Building „AgroBioTech" Research Centre ITMS 26220220180.

Project VEGA 1/0464/17 "Monitoring of the impact of ecological fuels obtained from agricultural production and additives in hydrocarbon fuels to technical and environmental performance of internal combustion engines used in agricultural and transport machines"

Project VEGA 1/0155/18 „Applied research of the use of ecological energy carriers in agricultural, forestry and transport technology."

Pproject APVV SK-PL-18-0041 „The development of scientific cooperation in the study of the effect of biofuels in road transport, including environmental impact."

## References

1. P. Lenďák, J. Jablonický, D. Uhrinová, J. Kosiba, J. Polerecký, *AMR, vol* **1059,** pp. 119–125, (2014)
2. J. Jablonický, Z. Tkáč, M. Majdan, D. Uhrinová, Ľ. Hujo, V. Vozárová. 2012 *Properties evaluation of biofuels and bio-lubricants* (Press Slovak University of Agriculture in Nitra, Nitra, 2012)
3. M. Králik, J. Jablonický, M. I. Nikolov, *Monitoring of NOx emissions at selected diesel engine* (Angel Kanchev University of Ruse, Ruse, 2015)
4. J. Čedík, M. Pexa, B. Peterka, M. Holubek, D. Mader, R. Pražan, *ATA,* **21**(4) (2018), pp. 130–135
5. I. Janoško, P. Kuchar, *Agron. Res.,* **16**(3) (2018), pp. 737–748
6. P. Dostál, J. Rozlivka, V. Kumbár, 2019. *ATA,* **(22)** 1, (2019) pp. 17-21
7. J. Čupera, P. Štěrba, 2010. *Diagnosis of motor vehicles I.* (Avid s.r.o. Brno, 2010)
8. N. V. Heeb, M. Zennegg, R. Haag, C. Seiler, P. Schmid, A. Wichser, A. Ulrich, P. Honegger, K. Zeyer, L. Emmeneggerer, Y. Zimmerli, J. Czerwinski M. Kasper, M. Mayer, *In Conference Proc. 15th ETH Conference on Combustion Generated Nanoparticles*. Zurich; Switzerland, Poster (2011)
9. F. Vlk F. *Automotive combustion engines* (VLK, Brno, 2003)
10. Y. Ulusoy, R. Arslan, C. Kaplan, *Energy Sources,* **31**(11) (2009), pp. 906–910
11. G. P. Hammond, S. Kallu, M. C. McManus, *Appl. Energy* **85**(6) (2018), pp. 506–515.
12. Regulation no. 24 of the Economic Commission for Europe of the United Nations (UN/ECE). Uniform provisions concerning: I. The approval of compression ignition (C.I.) engines with regard to the emission of visible pollutants; II. The approval of motor vehicles with regard to the installation of C.I. engines of an approved type; III.

The approval of motor vehicles equipped with C.I. engines with regard to the emission of visible pollutants by the engine; IV. The measurement of power of C.I. engines
13. M. P. Dorado, E. Ballesteros, J. M. Arnal, J. Gómez, F. J. López, *Fuel* **82**(11), (2003), pp. 1311–1315
14. Regulation no. 49 of the Economic Commission for Europe of the United Nations (UN/ECE). Emissions of C.I. and P.I. (NG and LPG) engines
15. Directive 2014/45/EU of the European Parliament and of the Council of 3 April 2014 on periodic roadworthiness tests for motor vehicles and their trailers and repealing Directive 2009/40/EC Text with EEA relevance
16. Law no. 106/ 2018 of 14 March 2018 on the operation of vehicles in road traffic and on the amendment of certain laws
17. ISO 11614:1999, (1999)
18. V. Rataj, *Preparing a scientific and professional text,* (SUA in Nitra, pp. 171, 2003)
19. B. F. Lin, J. H. Huang, D. Y. Huang, *Fuel* **88**(9) (2009), pp. 1779–1785
20. N. Y. Kado, P. A. Kuzmicky, National Renewable Energy Laboratory, p. 26. NREL/SR–510–31463, (2003)
21. J. Čedík, M. Pexa, B. Peterka, M. Holubek, D. Mader, R. Pražan, 2018. *Agron. Res.* **16** (S1) (2018), pp. 985–996
22. S. Puhan, N. Vedaraman, N., Sankaranarayanan, B. V. Bharat, Ram, *Renew. Energy* **30**(8) (2005), pp. 1269–1278
23. A. N. Ozsezen, M. Canakci, A. Turkcan, C. Sayin, *Fuel* **88**(4) (2009), pp. 629–636
24. P. K. Sahoo, L. M. Das, M. K. Babu, P. Arora, V. P. Singh, N. R. Kumar, T. S. Varyni, *Fuel* **88**(9) (2009), pp. 1698–1707
25. J. M. Luján, V. Bermúdez, B. Tormos, B. Pla, *Biomass Bioenerg.* **33**(6-7) (2009), pp. 948–956
26. M. J. Haas, K. M. Scott, T. L. Alleman, R. L. McCormick, *Energ. Fuel,* **15**(5) (2001), pp. 1207–1212

# Application of the Pareto front for risk control in the transport system

*Agnieszka* Sołtysiak[1,*], *Klaudiusz* Migawa[1]

[1]UTP University of Science and Technology, Faculty of Mechanical Engineering, 85-796 Bydgoszcz, Poland

**Abstract.** The article describes the developed model of controlling the process of means of transport operation, in which the choice of control strategy is carried out using non-deterministic methods. The model presented in the paper allows one to evaluate the quality of the transport system operation from the point of view of selected evaluation criteria: the risk of occurrence of undesired events and the availability of means of transport. The article presents a description of the method for determining the optimal (quasi-optimal) strategy for controlling the process of the use of means of transport taking into account the semi-Markov decision-making processes. The selection of the optimal (quasi-optimal) solution is carried out using the genetic algorithm and the simulated annealing algorithm. As a result of numerical calculations, for the criterion functions used, a set of quasi-optimal solutions is obtained in the form of the so-called Pareto front. This applies to the selection of possible decision variants, such a strategy for controlling the operation process, for which the functions constituting the evaluation criteria achieve values belonging to the Pareto-optimal solutions set.

## 1 Introduction

The issues discussed in the article concern the problem of controlling the process of operation of technical objects. In complex systems of the use of technical objects, the choice of rational control decisions from among the possible decision-making variants should be implemented using appropriate scientific methods, not only in an "intuitive" manner, based solely on the knowledge and experience of the system decision-makers. Due to the considerable complexity of the modeled processes and technical systems, it is necessary to use various types and mathematical tools that ensure effective implementation of the research and the analysis of the results obtained. Depending on the type of the discussed research problems, appropriate methods for determining optimal or quasi-optimal solutions are used, eg: [3, 6, 8, 9, 10, 11, 16]. The use of appropriate mathematical methods to control the operation process facilitates the selection of rational control decisions in a way that ensures correct and effective implementation of the tasks assigned to the system. One of the methods of assessing the possibility of proper implementation of the assigned task is to determine the values of the characteristics describing the safety and efficiency of the system of the use of technical objects.

The article describes the method for determining the optimal (quasi-optimal) strategy for controlling the operation of technical objects using decision-making semi-Markov processes, in which the evaluation criteria include the risk of undesired events and availability in the transport system operation. Moreover, paper [11] in which the selection of the optimal strategy was carried out using genetic algorithm, contains the semi-Markov decision-making model of availability control. On the other hand, paper [3] discusses the

[*] Corresponding author: agnieszka.soltysiak@utp.edu.pl

issues regarding the safety and risk evaluation, also analyzing the marine safety control model developed using decision-making semi-Markov processes as well as the Howard algorithm. Additionally, in paper [10] the semi-Markov model of risk evaluation in the transport system was developed. In this paper, two multi-criteria algorithms (genetic and simulated annealing) belonging to the group of non-deterministic methods were used to determine rational control strategies, the use of which makes it possible to obtain a result in the form of a set of optimal solutions in the Pareto sense (set of solutions belonging to the so-called Pareto Front). The application of these methods made obtaining simple and quick solutions in complex issues of multi-criteria optimization possible.

Genetic algorithm belongs to the group of methods in which the determination of the optimal solution is obtained as a result of determining successive solutions as random modifications of previous solutions dependent on them in a significant way. The basic assumption of using genetic algorithm to search for the optimal solution is the fact taken from the theory of evolution that the highest probability of modification concerns solutions with the highest degree of adaptation, determined by the value of the adaptation function (the function of the optimization task). The concept of a single-criterion genetic algorithm was developed by Holland in the 1960s [5]. The first genetic algorithm used to solve multi-criteria optimization problems was presented by Schaffer in 1985 and was called Vector Evaluated Genetic Algorithm (VEGA) [13]. The most important out of the numerous methods of this type include [7]: Multi-objective Genetic Algorithm (MOGA), Niched Pareto Genetic Algorithm (NPGA), Weight-Based Genetic Algorithm (WBGA), Random Weighted Genetic Algorithm (RWGA), Non-dominated Sorting Genetic Algorithm (NSGA), Strength Pareto Evolutionary Algorithm (SPEA), Pareto-Archived Evolution Strategy (PAES), Pareto Envelope-based Selection Algorithm (PESA).

The simulated annealing algorithm belongs to the group of heuristic methods in which the optimal solution is obtained as a result of searching for better and better solutions in subsequent iterations. Such algorithms are used to search for solutions close to optimal solutions in large optimization tasks, and the way they work is similar to the annealing process used in metallurgy. In the literature on the subject, many studies can be found concerning both theoretical description and examples of practical applications of the simulated annealing algorithm for searching for the optimal solution, eg: [1, 4, 12, 14, 15, 17]. The most important methods of this kind include: the MOSA multicriteria algorithm of simulated annealing with the acceptance criterion based on Pareto dominance and the AMOSA archived algorithm of simulated annealing.

In the case of risk related to the operation of technical systems, many methods are used to evaluate and analyze threats or levels of security, including qualitative, analytical, graphic, as well as quantitative methods. Although these methods allow to assess, control and reduce the risk values to acceptable levels, they do not take into account or only take into account, to a limited extent, the impact of significant parameters of the operation process of technical objects. There is no link between the risk assessment and the completion of the criteria for ensuring the correct implementation of the assigned tasks, eg taking into account the required level of availability, reliability and the value of economic indicators in the technical objects operation system. Such an approach requires taking into account, in addition to the risk, an additional criterion for evaluating the operation of the technical system. The paper presents the results of research of the decision model, the use of which allows to determine a rational strategy for controlling the operation process based on the evaluation of the risk of undesired events, while taking into account the requirements for technical objects for the implementation of assigned transport tasks.

## 2 Decision-making model for determining control strategies

Due to the random nature of factors affecting the process of operation of technical objects (eg. means of transport), stochastic processes are most often used for mathematical modeling of the operation process. Among the random processes, the Markov and semi-Markov processes were widely used in modeling the process of exploitation of technical objects, while in the case of issues related to the control of complex operation processes - Markov and semi-Markov decision-making processes.

Assuming that the analyzed model of the process of means of transport operation is a stochastic process $\{X(t): t \geq 0\}$ with finite number of process states $i \in S = \{1,2,\ldots,m\}$, then

$$D_i = \{d_i^{(1)}(t_n), d_i^{(2)}(t_n), \ldots, d_i^{(k)}(t_n)\} \tag{1}$$

means a set of all possible decisions that can be applied in the $i$-th state of the process at the moment $t_n$, where $d_i^{(k)}(t_n)$ means the $k$-th controlling decision made at the $i$-th process state at the moment $t_n$.

In the case where the optimization task consists of choosing the optimal strategy for controlling the operation process from among the acceptable strategies, then the $\delta$ strategy is to be understood as a sequence expressed by vectors, made up of decisions $d_i^{(k)}(t_n)$ made at moments $t_n$ of changes of states of the modeled technical object operation process

$$\delta = \{[d_1^{(k)}(t_n), d_2^{(k)}(t_n), \ldots, d_m^{(k)}(t_n)]: \; n = 0,1,2,\ldots\} \tag{2}$$

In the case when decisions made at successive states of the process do not depend on the moment $t_n$ at which they are taken, that is $d_i^{(k)}(t_n) = d_i^{(k)}$, the strategy $\delta$ is called the stationary strategy. Then the formula (2) takes the following form

$$\delta = [d_1^{(k)}, d_2^{(k)}, \ldots, d_m^{(k)}] \tag{3}$$

In order to determine the optimal control strategy (decision sequence) it is possible to apply decision-making semi-Markov processes. The decision-making semi-Markov process is a stochastic process $X(t)$, $t \geq 0$ the implementation of which depends on the decisions made at the initial process moment $t_0$ and in the moments of process changes $t_1$, $t_2$, …, $t_n$, …. . In the case of using semi-Markov decision-making processes, making the $k$-th control decision at the moment $t_n$ of the $i$-th state of the process means the selection of the $i$-th line of the process kernel from the set

$$\left\{Q_{ij}^{(k)}(t): t \geq 0, \;\; d_i^{(k)}(t_n) \in D_i, \;\; i,j \in S\right\} \tag{4}$$

where

$$Q_{ij}^{(k)}(t) = p_{ij}^{(k)} \cdot F_{ij}^{(k)}(t) \tag{5}$$

The selection of the $i$-th line of the process kernel determines the probabilistic mechanism of the process evolution in the time interval $<t_n, t_{n+1})$. This means that for the semi-Markov process, if the process state changes from any given one to the $i$-th (entry into the $i$-th state of the process) at the moment $t_n$, decision $d_i^{(k)}(t_n) \in D_i$ is made; moreover, according to the distribution $(p_{ij}^{(k)} : j \in S)$, the $j$-th process state is generated, transition into which takes place at the moment $t_{n+1}$. At the same time, according to the distribution defined by the distribution function $F_{ij}^{(k)}(t)$, the length of the time interval $<t_n, t_{n+1})$ for remaining at the $i$-th state of the process is generated, when the next state is $j$-th state.

In the paper, for the developed decision-making model two criteria functions were used: unit risk of occurrence of an undesired event $r(\delta)$ [PLN/h] and availability of a single technical object $G(\delta)$:

$$f_{C_1}(\delta)=r(\delta)=\sum_{i\in S_N}c_i(\delta)\cdot p_i^*(\delta)=\frac{\sum_{i\in S_N}c_i(\delta)\cdot\pi_i\cdot\Theta_i(\delta)}{\sum_i\pi_i\cdot\Theta_i(\delta)} \tag{6}$$

$$f_{C_2}(\delta)=G(\delta)=\sum_{i\in S_G}p_i^*(\delta)=\frac{\sum_{i\in S_G}\pi_i\cdot\Theta_i(\delta)}{\sum_{i\in S}\pi_i\cdot\Theta_i(\delta)} \tag{7}$$

where:

$S_N \subset S$ – set of undesirable states of the modeled operation process,

$S_G \subset S$ – set of availability states of the modeled operation process,

$c_i(\delta)$ – unit revenue generated at process states $X(t)$,

$p_i^*(\delta)$ – border probabilities for remaining at the states of the analyzed process $X(t)$, determined on the basis of border theorem for semi-Markov processes [2]

$$p_i^*(\delta)=\frac{\pi_i\cdot\Theta_i(\delta)}{\sum_{i\in S}\pi_i\cdot\Theta_i(\delta)} \tag{8}$$

where:

$\Theta_i(\delta)$ – mean values for unconditional time lengths of process states,

$\pi_i$ – probabilities of stationary distribution of the embedded Markov chain completing the system of linear equations

$$\sum_{i\in S}\pi_i\cdot p_{ij}=\pi_j,\ \ j\in S,\ \ \sum_{i\in S}\pi_i=1 \tag{9}$$

where:

$p_{ij}$ – conditional probabilities of transition from state $i$ to state $j$, according to the dependence:

$$p_{ij}=\lim_{t\to\infty}p_{ij}(t) \tag{10}$$

$$p_{ij}(t)=P\{X(t)=j|X(0)=i\} \tag{11}$$

In the semi-Markov decision-making model, the choice of rational control strategy $\delta$ referred to as optimal strategy $\delta^*$ applies to the situation when the function (functions) constituting the criterion for selecting the optimal strategy assumes the extreme value (minimum or maximum)

$$f_C(\delta^*)=\min_\delta[f_C(\delta)]\ \ or\ \ f_C(\delta^*)=\max_\delta[f_C(\delta)] \tag{12}$$

then, in the decision model presented in the paper, the selection of the rational (optimal) strategy $\delta^*$ is made on the basis of the following criteria:

$$r(\delta^*)=\min_\delta[r(\delta)] \tag{13}$$

$$G(\delta^*)=\max_\delta[G(\delta)] \tag{14}$$

In the developed decision-making model, the genetic algorithm and the simulated annealing algorithm were used to select the strategy to control the process of means of transport operation. In the case of using this type of mathematical tools to determine the strategy of optimal $\delta^*$ control of the operation process, the following assumptions should be made:

- the examined model of the process of means of transport operation is an $m$-state stochastic process,

- at each state of the operating model, you can use one of the two possible decisions $D = \{0,1\}$,
- if the decisions are marked as 0 and 1, then the number of possible control strategies for the $m$-state model of the operation process amounts to $2^m$,
- set of control strategies is a set of the function $\delta : S \rightarrow D$.

Based on the above assumptions, every possible control strategy can be presented as a $m$-positioned sequence consisting of 0 and 1, then a typical control strategy for a 19-state model is defined as follows: $\delta = [1,0,0,1,1,0,1,0,1,1,1,0,1,1,0,1,0,0,1]$.

## 3 Defining rational control strategy

The model of the operation process was built on the basis of the analysis of the state space as well as operational events concerning city buses operating in the analyzed transport system. Due to the analyzed evaluation criteria: the risk of occurrence of undesired events and the availability of technical objects, based on the identification of the running multi-state process of operation of technical objects, significant operational states of this process were determined and possible transitions between the distinguished states. The following list contains the states of the modeled operation process and a probabilities matrix of transitions between these states:
**1** – carrying out of transport task, **2** – supply, **3** – servicing on day of operation OC, **4** – periodical technical service OT, **5** – stoppage at the depot yard, **6** – stoppage at route due to damage, **7** – diagnostics performed by emergency service PT, **8** – repair by emergency service PT without missing a ride, **9** – repair by emergency service PT with a missed ride, **10** – waiting for completion of the transport task after repair by technical service PT, **11** - emergency exit, **12** - diagnostics before repair at service station SO, **13** - repair at service station SO, **14** - diagnostics after repair at service station SO, **15** - stoppage caused by an accident or collision, **16** - intervention and rescue operation after an accident or collision, **17** - repair after an accident or collision, **18** - vehicle replacement, **19** - diagnostics after post-accident repair.

$$P = \begin{bmatrix}
0 & p_{1,2} & 0 & 0 & 0 & p_{1,6} & 0 & 0 & 0 & 0 & 0 & 0 & 0 & 0 & p_{1,15} & 0 & 0 & 0 & 0 \\
0 & 0 & p_{2,3} & p_{2,4} & 0 & 0 & 0 & 0 & 0 & 0 & 0 & 0 & 0 & 0 & 0 & 0 & 0 & 0 & 0 \\
0 & 0 & 0 & 0 & p_{3,5} & 0 & 0 & 0 & 0 & 0 & 0 & 0 & p_{3,13} & 0 & 0 & 0 & 0 & 0 & 0 \\
0 & 0 & 0 & 0 & p_{4,5} & 0 & 0 & 0 & 0 & 0 & 0 & 0 & p_{4,13} & 0 & 0 & 0 & 0 & 0 & 0 \\
p_{5,1} & 0 & 0 & 0 & 0 & 0 & 0 & 0 & 0 & 0 & 0 & 0 & 0 & 0 & 0 & 0 & 0 & 0 & 0 \\
0 & 0 & 0 & 0 & 0 & 0 & p_{6,7} & 0 & 0 & 0 & p_{6,11} & 0 & 0 & 0 & 0 & 0 & 0 & 0 & 0 \\
0 & 0 & 0 & 0 & 0 & 0 & 0 & p_{7,8} & p_{7,9} & 0 & p_{7,11} & 0 & 0 & 0 & 0 & 0 & 0 & 0 & 0 \\
p_{8,1} & 0 & 0 & 0 & 0 & 0 & 0 & 0 & 0 & 0 & 0 & 0 & 0 & 0 & 0 & 0 & 0 & 0 & 0 \\
p_{9,1} & 0 & 0 & 0 & 0 & 0 & 0 & 0 & 0 & p_{9,10} & 0 & 0 & 0 & 0 & 0 & 0 & 0 & 0 & 0 \\
p_{10,1} & 0 & 0 & 0 & 0 & 0 & 0 & 0 & 0 & 0 & 0 & 0 & 0 & 0 & 0 & 0 & 0 & 0 & 0 \\
0 & 0 & 0 & 0 & 0 & 0 & 0 & 0 & 0 & 0 & 0 & p_{11,12} & p_{11,13} & 0 & 0 & 0 & 0 & 0 & 0 \\
0 & 0 & 0 & 0 & 0 & 0 & 0 & 0 & 0 & 0 & 0 & 0 & p_{12,13} & 0 & 0 & 0 & 0 & 0 & 0 \\
0 & p_{13,2} & 0 & 0 & 0 & 0 & 0 & 0 & 0 & 0 & 0 & 0 & 0 & p_{13,14} & 0 & 0 & 0 & 0 & 0 \\
0 & p_{14,2} & 0 & 0 & 0 & 0 & 0 & 0 & 0 & 0 & 0 & 0 & 0 & 0 & 0 & 0 & 0 & 0 & 0 \\
0 & 0 & 0 & 0 & 0 & 0 & 0 & 0 & 0 & 0 & 0 & 0 & 0 & 0 & 0 & p_{15,16} & 0 & 0 & 0 \\
0 & 0 & 0 & 0 & 0 & 0 & 0 & 0 & 0 & 0 & 0 & 0 & 0 & 0 & 0 & 0 & p_{16,17} & 0 & p_{16,19} \\
0 & 0 & 0 & 0 & 0 & 0 & 0 & 0 & 0 & 0 & 0 & 0 & 0 & 0 & 0 & 0 & 0 & p_{17,18} & 0 \\
0 & p_{18,2} & 0 & 0 & 0 & 0 & 0 & 0 & 0 & 0 & 0 & 0 & 0 & 0 & 0 & 0 & 0 & 0 & 0 \\
0 & p_{19,2} & 0 & 0 & 0 & 0 & 0 & 0 & 0 & 0 & 0 & 0 & 0 & 0 & 0 & 0 & 0 & 0 & 0
\end{bmatrix}$$

Controlling the process of city buses operation is possible as a result of making appropriate decisions in the decision-making states of the process. For the model of the urban bus operation process in question, based on operational data, the values of probability transition matrix elements $P$ were estimated, possible decisions at the decision-making states of the process (Table 1) as well as unconditional dates and unit revenues generated at the states of the analyzed process were determined (Table 2).

**Table 1.** Decision at decision-making states of the analysed process.

| Process state | Decision "0" | Decision "1" |
|---|---|---|
| 1 | Route with "light" conditions for completing the transport task | Route with "heavy" conditions for completing the transport task |
| 3 | OC servicing – "standard" | OC servicing – "intense" |
| 4 | OT servicing – "standard" | OT servicing – "intense" |
| 7 | PT diagnostics – "standard" | PT diagnostics – "intense" |
| 8 | PT repair without missing a ride – "standard" | PT repair without missing a ride – "intense" |
| 9 | PT repair with a missed ride – "standard" | PT repair with a missed ride – "intense" |
| 11 | Emergency exit – no assistance | Emergency exit – towing |
| 12 | Diagnostics before repair at SO – "standard" | Diagnostics before repair at SO – "intense" |
| 13 | Repair at SO – "standard" | Repair at SO – "intense" |
| 14 | Post-repair diagnostics at SO – "standard" | Post-repair diagnostics at SO – "intense" |
| 16 | Post-collision intervention – "standard" | Post collision intervention – "intense" |
| 17 | Post-repair or post-collision repair – "standard" | Post-repair or post-collision repair – "intense" |
| 18 | Post-accident repair diagnostics – "standard" | Post-accident repair diagnostics – "intense" |

**Table 2.** Mean lengths of time and unit revenues at the states of the analyzed process depending on decision.

| Process state | $\Theta_i^{(0)}$ [h] | $\Theta_i^{(1)}$ [h] | $c_i^{(0)}$ [PLN/h] | $c_i^{(1)}$ [PLN/h] |
|---|---|---|---|---|
| 1 | 8.9 | 7.8 | 28.5 | 33.5 |
| 2 | 0.1 | 0.1 | -23 | -23 |
| 3 | 0.15 | 0.11 | -215 | -250 |
| 4 | 3.9 | 2.9 | -210 | -235 |
| 5 | 5.7 | 5.7 | -6.8 | -6.8 |
| 6 | 0.055 | 0.055 | -24.5 | -24.5 |
| 7 | 0.25 | 0.19 | -55 | -70 |
| 8 | 0.10 | 0.06 | -70 | -97 |
| 9 | 0.55 | 0.45 | -102 | -132 |
| 11 | 0.45 | 0.45 | -24.5 | -24.5 |
| 12 | 0.41 | 0.35 | -110 | -140 |
| 13 | 0.42 | 0.38 | -83.5 | -110 |
| 14 | 3.4 | 2.7 | -117 | -136 |
| 15 | 0.42 | 0.38 | -83.5 | -110 |
| 16 | 0.4 | 0.4 | -245 | -245 |

| | | | | |
|---|---|---|---|---|
| 17 | 2.2 | 1.7 | -400 | -540 |
| 18 | 120 | 100 | -100 | -200 |
| 19 | 0.8 | 0.6 | -55 | -84 |

The presented example has been developed on the basis of operational data obtained from the tests of the actual system of operation of means of transport (city transport buses). 160 city buses are used in the analyzed transport system, and service and repair processes are carried out at technical support stations and by technical emergency units.

In the presented model of the operation process of city buses, states between 6 and 19 are undesirable states (they influence the value of the risk of occurrence of undesirable events), while states 1, 5, 8 and 10 are states of the availability of the technical object to carry out the transport tasks. To add, the calculations were made using an operational computer program using the simulated annealing algorithm and the genetic algorithm. The computer program was developed by R Development Core Team (2018). R: A language and environment for statistical computing. R Foundation for Statistical Computing, Vienna, Austria. ISBN 3-900051-07-0. On the basis of the above calculations, the optimal criteria (quasi-optimal) control strategies for the operation process carried out in the analyzed transport system were determined for the adopted criteria. The results of calculations are shown in Figures 1 and 2 and in Tables 3 and 4.

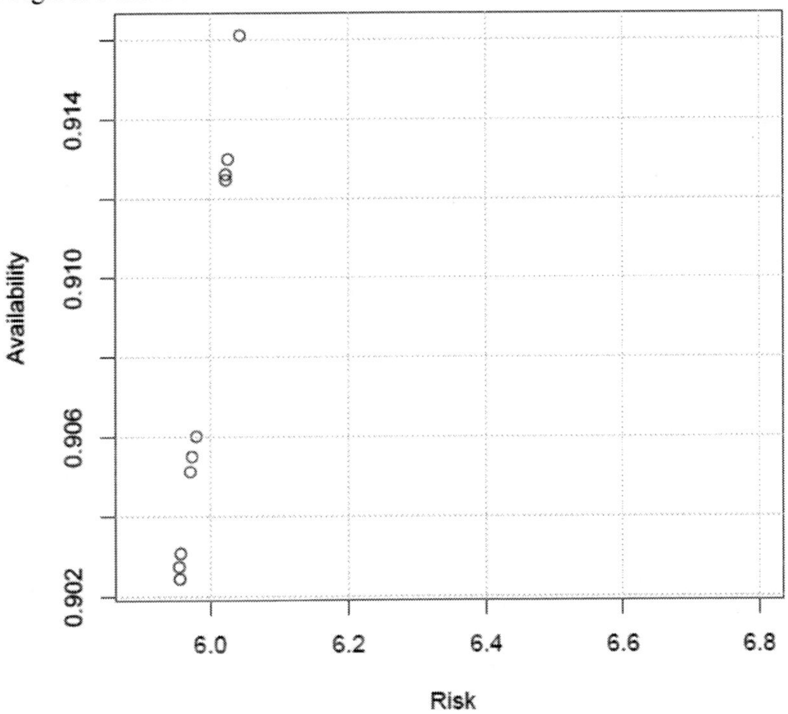


**Fig. 1.** The Pareto front of optimal solutions determined using simulated annealing algorithm [Availability/Risk].

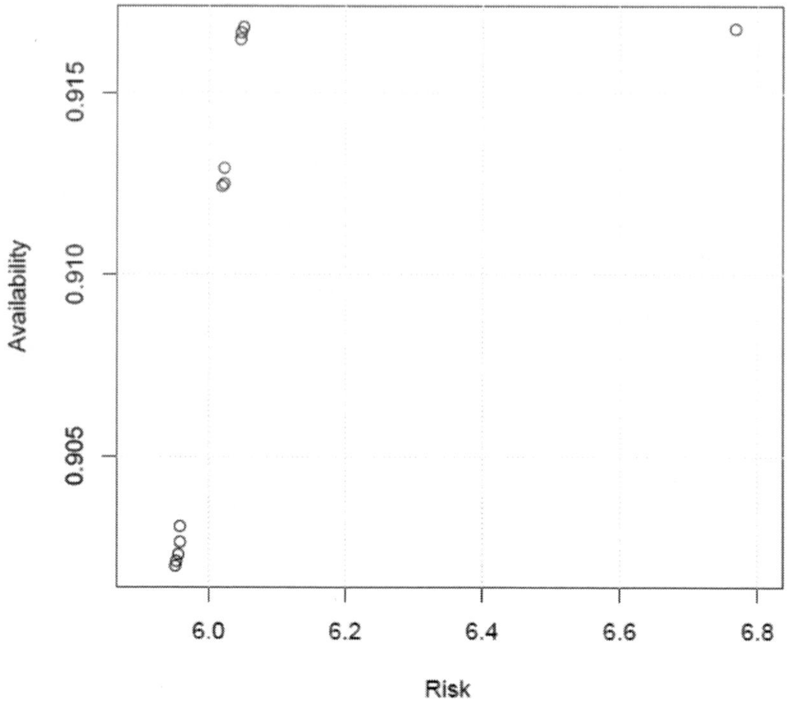


**Fig. 2.** The Pareto front of optimal solutions determined using genetic algorithm [Availability/Risk].

**Table 3.** Optimal control strategies $\delta^*$ and values of criterion functions determined using simulated annealing algorithm.

| **Strategy $\delta^*$** | **$r(\delta^*)$ [PLN/h]** | **$G(\delta^*)$** |
|---|---|---|
| [0,0,0,0,0,0,1,1,1,1,1,1,0,0,1,0,0,0,0] | 5.9553 | 0.9027 |
| [0,0,0,1,1,0,0,1,1,0,0,0,0,0,1,1,0,1,0] | 5.9562 | 0.9024 |
| [0,1,0,1,1,0,1,1,0,0,1,1,0,1,1,0,0,0,0] | 5.9575 | 0.9031 |
| [0,0,1,0,1,0,0,0,0,0,1,0,0,0,0,0,0,0,0] | 5.9716 | 0.9051 |
| [0,1,1,1,0,1,0,0,1,1,0,1,0,0,1,0,0,0,0] | 5.9742 | 0.9055 |
| [0,1,1,1,0,0,1,0,1,0,0,0,0,0,1,1,0,1,0] | 5.9804 | 0.9060 |
| [0,0,0,0,1,0,0,1,0,0,1,0,1,1,1,1,0,0,0] | 6.0229 | 0.9126 |
| [0,0,0,0,1,0,0,0,1,1,0,0,1,0,0,1,0,1,0] | 6.0230 | 0.9125 |
| [0,1,0,1,0,1,0,0,1,0,0,1,1,1,0,1,0,0,0] | 6.0260 | 0.9130 |
| [0,1,1,0,0,0,1,1,1,1,0,0,1,0,0,0,0,0,0] | 6.0435 | 0.9161 |

**Table 4.** Optimal control strategies $\delta^*$ and values of criterion functions determined using genetic algorithm.

| **Strategy $\delta^*$** | **$r(\delta^*)$ [PLN/h]** | **$G(\delta^*)$** |
|---|---|---|
| [0,0,0,0,1,0,0,1,0,1,0,1,0,0,1,0,0,1,0] | 5.9507 | 0.9020 |
| [0,1,0,0,0,1,0,0,0,1,0,0,0,1,1,0,0,1,0] | 5.9521 | 0.9021 |

| | | |
|---|---|---|
| [0,0,0,0,0,1,0,1,1,0,0,0,0,1,1,1,0,1,0] | 5.9553 | 0.9023 |
| [0,1,0,1,0,1,1,1,0,0,1,1,0,1,0,0,0,0,0] | 5.9575 | 0.9031 |
| [0,1,0,0,0,1,1,0,1,0,0,1,0,0,0,1,0,0,0] | 5.9578 | 0.9026 |
| [0,1,0,0,1,1,0,1,0,1,1,1,1,0,1,0,0,0,0] | 6.0193 | 0.9124 |
| [0,1,0,1,1,0,0,1,1,0,0,0,1,1,0,0,0,0,0] | 6.0225 | 0.9129 |
| [0,0,0,0,1,1,0,1,0,0,0,0,1,1,1,1,0,1,0] | 6.0226 | 0.9125 |
| [0,0,1,1,1,1,1,0,1,0,0,1,1,1,0,0,1,0] | 6.0462 | 0.9165 |
| [0,1,1,1,0,0,1,1,1,1,0,1,1,1,1,0,0,0,0] | 6.0470 | 0.9166 |
| [0,1,1,1,0,0,1,1,1,0,1,1,1,1,0,1,0,0,0] | 6.0505 | 0.9168 |
| [0,1,1,1,0,0,1,1,1,0,1,0,1,1,0,1,1,0,0] | 6.7679 | 0.9168 |

## 4 Summary

Paper [11] discusses the model of the process of operation of means of transport, the use of which enables the designation of control strategies with the help of a single criterion (availability of technical objects). The method presented in the article, however, applies to multi-criteria analysis (eg. using two evaluation criteria such as risk and availability). On the basis of the results obtained, it can be noticed that increasing the availability of means of transport is associated with an increased risk of undesired events. This may be due to the fact that in the developed model the increase in the level of availability of technical objects does not reduce the probability of undesired events, but only causes the necessity of incurring additional costs related to the removal of the effects of these events. This is due to the need to provide additional actions and measures for the treatment of technical objects. This is accomplished by increasing the intensity of service and repairs performed, e.g. using more efficient equipment and tools, and a larger number of employees.

Analyzing the obtained results, it can be noticed that both in the case of simulated annealing algorithm and genetic algorithm, the values of the criterion functions are included in similar ranges for the designated control strategy sets. The risk values for undesired events are relevant from 5.9553 to 6.0435[PLN/h], with the availability of technical objects ranging from 0.9027 to 0.9161 for simulated annealing algorithm and from 5.9507 to 6.0505 [PLN/h], with the availability of technical objects ranging from 0.9020 to 0.9168 for genetic algorithm. Only in the case of genetic algorithm, for one of the designated control strategies a higher value of the risk of undesired events amounted to 6.7679 [PLN/h], however, without increasing the level of availability of technical objects.

For the determined values of the criterion functions, the corresponding rational control strategies have been developed as sets of optimal solutions in the Pareto sense. The optimal set in the Pareto sense is a set of non-dominated solutions of the entire permissible search space. Optimal solutions in the Pareto sense create the so-called Pareto front. Then, based on the results obtained, one may select a single solution from among the set of optimal solutions located on the Pareto front. Such selection is usually made by the decider (team of deciders) of the system, based on additional premises regarding both the specific decision-making situation and the current conditions in which the operating system operates.

Depending on the demands, the developed algorithms (simulated annealing or genetic), along with the decision-making model of the operation process can be applied to the mathematical formulation and solving of a wide range of problems related to the control of complex technical systems. This applies mainly to the economic analysis, risk and security management as well as to the availability and reliability of the technical objects under operation.

The model presented in the article is a partial result of the conducted research the aim of which is to develop a comprehensive method of controlling the operation of means of transport.

## References

1. P. Czyżak, A. Jaszkiewicz, *Pareto simulated annealing - A metaheuristic technique for multiple-objective combinatorial optimization* (J. Multi-Criteria Decision Anal., 1998)
2. F. Grabski, *Semi-Markov processes: Applications in system reliability and maintenance*. (Amsterdam, Elsevier, 2014)
3. F. Grabski, *Analiza ryzyka w decyzyjnych semi-markowskich modelach procesu eksploatacji. Risk analysis in decision-making semi-Markov models of operation process,* (XXXVIII Zimowa Szkoła Niezawodności. 38th Winter Reliability School, Szczyrk, 2010)
4. M. Hapke, A. Jaszkiewicz, R. Słowiński, *Pareto simulated annealing for fuzzy multi-objective combinatorial optimization.* (J. Heuristics, 2000).
5. J. H. Holland, *Adaptation in natural and artificial systems* (Ann Arbor, University of Michigan Press, 1975)
6. L. Knopik, K. Migawa, A. Wdzięczny, *Profit optimization in operation systems*. (Polish Maritime Research, 2016)
7. A. Konak, D. W. Coit, A. E. Smith, *Multi-objective optimization using genetic algorithms: A tutorial* (Reliability Engineering and System Safety, Elsevier, 2006)
8. V. G. Kulkarni, *Modeling and analysis of stochastic systems* (New York, Chapman & Hall, 1995)
9. K. W. Lee, *Stochastic models for random-request availability* (IEEE Trans, Reliability, 2000)
10. K. Migawa, L. Knopik, A. Sołtysiak, P. Kolber, *The method of risk assessment in transport system* (Engineering Mechanics, Czech Republic, 2017)
11. K. Migawa, L. Knopik, S. Wawrzyniak, *Application of genetic algorithm to control the availability of technical systems* (Engineering Mechanics, Czech Republic, 2016)
12. D. K. Nam, C. Park, *Multiobjective simulated annealing: A comparative study to evolutionary algorithms* (Int. J. Fuzzy Syst., 2000)
13. J. D. Schaffer, *Multiple objective optimization with vector evaluated genetic algorithms* (In: Proceedings of the international conference on genetic algorithm and their applications, 1985)
14. B. Suman, P. Kumar, *A survey of simulated annealing as a tool for single and multiobjective optimization* (Journal of the Operational Research Society. 2006)
15. A. Suppapitnarm, K. A. Seffen, G. T. Parks, P. Clarkson, *A simulated annealing algorithm for multiobjective optimization* (Eng. Opt., 2000)
16. M. Szubartowski, K. Migawa, A. Neubauer, L. Knopik, *Method of determining optimal control strategy* (58th International Conference of Machine Design Departments, ICMD, 2017)
17. E. L. Ulungu, J. Teghaem, P. Fortemps, D. Tuyttens, *MOSA method: A tool for solving multiobjective combinatorial decision problems*. (J. Multi-Criteri Decision Anal., 1999)

# Exploitation of Car Turbochargers - Selected Problems

*Hubert* Wojciechowski[1,*], *Bogdan* Żółtowski[2]

[1]University of Science and Technology in Bydgoszcz, Faculty of Mechanical Engineering, Al. prof. S. Kaliskiego 7, 85-796 Bydgoszcz, Poland
[2]UTH Technical and Commercial University, Ul. Jutrzenki 135, 02-231 Warsaw, Poland

**Abstract.** In order to increase the filling of the cylinder in the engine, a so-called charge is applied, which consists in increasing the amount of air input under increased pressure and at an unchanged temperature. This increases the density of that air and increases the mass in the engine cylinder. In most cars, the charging is carried out by an exhaust gas turbocharger or a mechanical compressor. This paper presents the issues that concern the failure of turbochargers resulting from their balance by weight reduction method. Selected aspects concerning consequences of incorrect balancing and methods of reducing unbalanced rotor mass. A key part of the work includes a proposal to balance turbochargers of automotive vehicles using the "weight-adjustment"method. In this scope, preliminary studies were presente and conclusions observed during their implementation were described.

## 1 Introduction

Modern internal combustion engines are characterized by the downsizing (decrease in engine capacity, while maintaining high engine power) and aim to achieve the best overall efficiency [4].

These trends can be met by using mainly a turbo-charging system in the drive unit. The principle of the engine charging system is relatively simple and consists in injecting more air into the cylinder in order to burn the petrol-blend more efficiently. The turbocharger or compressor is responsible for this process. In this way, the engine produces more power, which improves vehicle performance.

Car companies use mechanical turbocharging in their models - compressors, as well as turbocharging. The second option is much more commonly used because of its universality. They are installed in compression ignition engines, spark ignition engines and hybrid cars. The turbocharger is the basic and at the same time the main element in the system of turbocharging [1,3,8]. It consists of three basic elements [3]: centre hull, in which there is a bearing turbocharger rotor, turbine hull and compressor hull.

The basis of the efficiency of the turbines is the fact that they are able to pump a very large amount of air into the cylinder, and in addition they are powered by exhaust gas energy, which is not used in any way in naturally aspirated engines. The turbo-charging system has a significant influence on performance, so a noticeable power decrease can be one of the symptoms of turbocharger damage, especially if the problem is mainly related to the rotational range in which the turbine is most active. The technical sources are usually rotor damages, bearing damage and, in the case of turbines with variable geometry, hydraulic steering wheels.

* Corresponding author: wojciechowskihubert@interia.pl

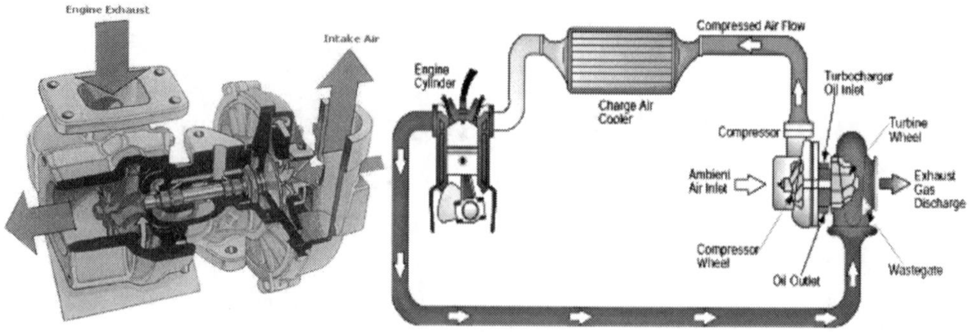


**Fig. 1.** Working principle of the turbocharger [2].

The turbocharger principle is based on the exhaust gas drive of the turbine rotor. The turbine is connected by a common shaft with a compression wheel, on which the air is aspirated. The compressed air is supplied to the combustion chamber of the engine cylinder under pressure - Fig. 1. To increase the efficiency of the turbocharger, an air cooler (intercooler) is used. Cooled air increases its density, which results an increase in the mass delivered to the combustion chamber.

Publications and scientific articles in this area of research discuss outside the typical operational defects of turbochargers do not focus on their faults, which are result from the methods of turbo-charging equipment regeneration [10,12]. Despite the huge popularity of engine turbo-charging systems for modern motor vehicles, manufacturers of turbochargers do not reveal the materials from which given components are constructed and do not provide information about theirs mechanical and physical properties [4]. Defects of the turbocharger components have a negative impact on the emission of harmful exhaust gases to the environment [13,15]. Description and modelling of the dynamic state, methods of diagnosing the state as well as the application of modern information techniques in turbochargers' research is so far hardly recognized and requires immediate improvement [14,16,17]. Literature describing the dependence of the method of regeneration and mainly balancing the turbocharger components -on the operation of the turbo-charging system is still very insignificant in this field. Existing publications so far describe mainly the balancing of turbochargers by means of a reduction method - **taking off unbalanced mass** and focusing mainly on damage to turbochargers resulting from improper operation of the turbo-charging device. Examples of this are the following articles: Balancing of turbocharger rotor [10], Damage analysis of turbocharger compressors [12], Analysis of the impact of operating conditions on the technical condition of turbocharger turbochargers of combustion engines [11]. Under heading [1] engine turbocharging, general information can be found on, inter alia, balancing turbochargers using the' mass capturing' method.

This article presents the main causes of turbocharger damage, emissions in motor vehicles. The present methods of verifying the condition of charging devices have also been described and an **innovative method of balancing turbochargers** by **"adding mass"** has been proposed. For the turbocharger regeneration, apart from the correct replacement of worn parts and the cleaning of other parts, the most important thing is the precise dynamic balancing of the rotating parts of the turbine.

## 2 Diagnostic tester - verification of actual parameters

In order to eliminate the undesirable condition of the turbocharger such as its failure, it is necessary to correctly identify and evaluate in detail the technical condition of the turbocharger by available ways. The simplest and at the same time lowest effective method of diagnosing turbocharger is organoleptic. Turbo-charging devices are evaluated with human senses. An oiled turbocharger may indicate a broken housing, damaged seals or a worn out suction section of the turbocharger. The organoleptics make it possible to hear disturbing sounds from the device. The damaged turbocharger is characterized by rustling, whistle. This can be indicated damaged, broken turbocharger rotor and a clogged cable between the inlet and intake manifold [1,3,7].

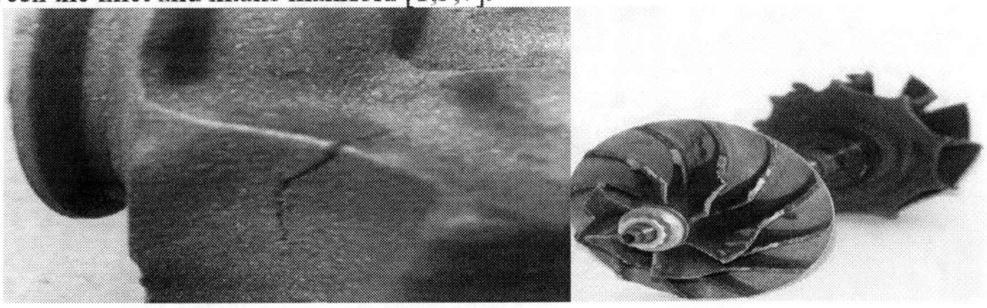

**Fig. 2.** Fracture of the outer part of the housing.

The colour of the exhaust gas coming out of the exhaust system may also indicate the condition of turbo- charging device. The blue or black colour of the exhaust gases can indicate a turbocharger failure. This defect may indicate a cracked compression wheel or a defective oil system in the turbocharger. The presented method is one of the simplest, but in case of its credibility it becomes questionable because it does not ensure accuracy and insight in its assessment [10].

The parameter that characterizes turbo- charge system and can indicate its condition is the pressure of turbocharging. The vehicle directive that has been drawn up stipulates that cars manufactured after 2000 must have 16-pin diagnostic connector. This makes it easier to diagnose any driver emissions in a vehicle and the reading of operating parameters in any time unit.

The key information in the turbo-charging circuit is the actual data, which is stored by signals from sensors in the circuit. The evaluation of the above parameters is possible through additional functions, which are used as applications in diagnostic testers. The turbo-charge pressure is checked here depending on [5]:

- rotor speed,
- compressed air pressure,
- exhaust gas temperatures,
- the blade angle of inclination in the variable geometry system.

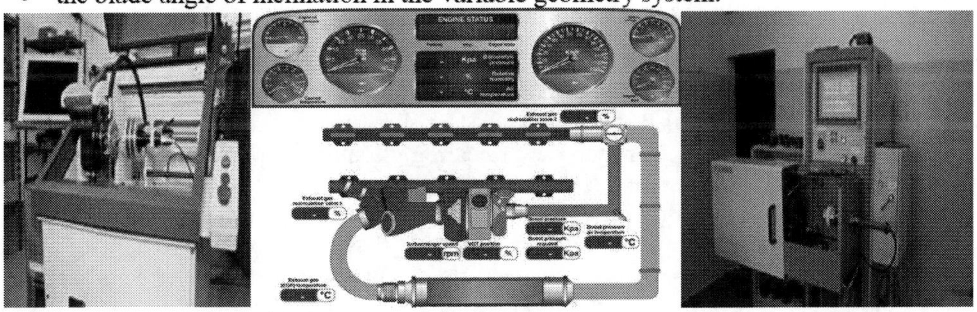

**Fig. 3.** DASHBOARD application in IDC5 software [5].

One of the main reasons of failure of turbochargers is unbalance of its rotating parts. To measure the balancing quality of the turbocharger's rotor we use two types of devices:

- quick-rotating turbocharger core correction machines (fig.4),
- slow rotating devices for balancing the turbocharger shaft (Fig. 5).

**Fig.4.** Core correction machine.

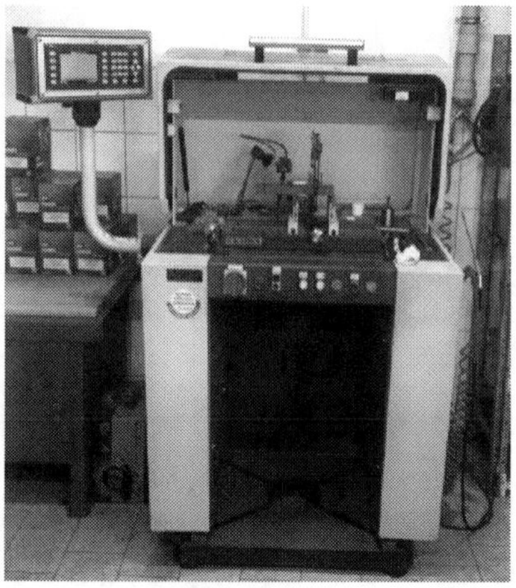

**Fig.5.** Rotor balancing machine.

The presented devices verify turbochargers in the context of unbalance measurements of rotor masses and correct the balance of turbine wheel and compression wheel with turbocharger shaft. Unbalanced weight is reduced with an accuracy of 0,01 mg/mm2 and the unbalance tolerance value is assigned to a given type of turbocharger and stored in the internal memory of the device. Unquestionably, the above method is one of the most accurate, precise methods of verifying turbo-charging devices.

## 3 Methods of identifying the condition of turbochargers

Due to higher and higher operating speeds of turbochargers, even a minor unbalance causes an increase of vibrations, which in consequences may lead to a quicker damage of the whole component. The effect of its use is subsequent vibrations of rotors, causing much faster wear of bearings, rings and other parts of turbochargers.

Balancing is the process of checking the distribution of the rotor mass and correcting it, if necessary, until residual unbalance is obtained within the set tolerances. The aim is to achieve an even weight distribution of the rotor with respect to its geometric axis of rotation. In theory, the unbalance of rotating elements can be divided into three basic types: static, momentary and dynamic. In practice, we are dealing with all of them simultaneously, but the division applied makes it easier to understand the essence of the phenomena taking place and to choose a rational method of limiting their negative results.

### 3.1 Dynamic balancing of the rotor by the 'mass reduction' method

The main reason for damage and failure of turbo-charging devices is unbalance of turbocharger's rotor. Rotor unbalance creates excessive vibrations which are transmitted to the machine body, which leads to damage to the roller bearings. A characteristic feature of the rotor unbalance is a regular increase in bearing vibration amplitude, which depends on the rotational speed of the rotor of the turbo-charging device. These are always harmonic vibrations, the frequency of which is equal to the frequency of rotor rotation. Rotational speed of turbocharger rotor oscillates around 200 000 rpm, so the problem of natural frequency is particularly important. The absolute condition of the turbocharger's usefulness is not showing a tendency to any vibrations, whether the drive unit rotates or not [2,6,16].

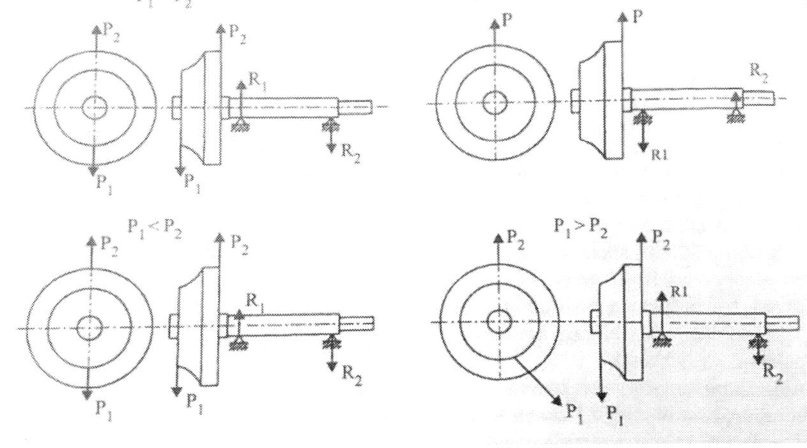


**Fig.7.** Illustrates the dynamic balance of the turbocharger's rotor [1].

The force pair P1 and P2 as shown in Fig. 7 is in the same mode of working as the axis of the rotor. On a bearing-mounted balancing station, the reactions R1 and R2 are generated, which are in the same position as the above forces. They have equal values but opposite directions (R1=-R2). The location of the rotor, as well as the value of unbalanced mass, is used to verify the place and mass of the material, which has to be removed for proper dynamic balancing of turbocharger rotor [1].

The turbocharger of vehicle is balanced in two stages:

- stage I - Balancing the shaft with the turbine wheel and the compression wheel with nuts and spacers and seals (Fig. 8),
- stage II - balancing of the bearing-mounted turbocharger rotor in the compressor housing with oil system connected (Fig. 9).

**Fig. 8.** Balancing the turbocharger shaft.

**Fig. 9.** Balancing the core of the turbocharger.

This type of balancing is the result of the recommended stricter requirements imposed by companies manufacturing turbo-charging devices. Correction of unbalance of turbochargers by "reduction of mass" certainly affects the structure of the balanced element, because excessive removal of material from the compression wheel or turbine wheel make them weak, which often causes cracking of these elements.

The following figures (10 and 11) show the points on the compression wheel from which the mass hass been reduced and the resulting wheel crack at the point where the unbalanced mass was gathered.

**Fig.10.** Mass reduction points on the compression wheel.

**Fig.11.** Cracked compression wheel.

### 3.2 Dynamic balancing of the rotor by the 'mass addition' method

Most of the damage to turbochargers is caused by breakage of the compression wheel; turbine; and also by breakage of the turbocharger shaft. Such defects are caused by improper exploitation of turbocharged combustion engines and incorrect regeneration of turbocharger components. Incorrect balancing of basic components has a negative impact on the construction and mechanical properties of turbocharger components. In order to prevent damage to materials which result from improper regeneration of turbocharger components, innovative tests are carried out in order to minimize the above problem. The method of 'mass addition' is carried out in two stages:

- first stage concerns the correction of the unbalanced mass of the turbocharger rotor including sealants and spacer nuts,
- second stage is based on balancing the rotating element (core of turbocharger) with bearings in the body of the device.

Two types of devices were used to perform these tests:

- Schenck TB Comfort slow-running balancing machine for the first stage of balancing,
- the second stage was carried out on a high-speed Schenck MBRS 110 machine.

Disassembly of balanced components and input of data of turbochargers into the system is the identical as in the traditional method using balancers for rotors and cores of turbochargers. The difference is in the way of correcting unbalanced masses of rotating parts. The process of unbalanced mass correction was carried out by using a special thermo-resistant mass, which is designed, among other things, for the removal of exhaust gases and liquids whose temperature exceeds 1000°C. Characteristic feature of this substance is a high adhesion coefficient and short time of mass fixation. During the reading of the value of the unbalanced mass of the rotating element, the heat-resistant mass is applied to the outside of the turbine wheel and compressed via an insulin syringe. Before application, the rotor surfaces were thoroughly cleaned, and the amount of mass applied was verified on an electronic scale with an accuracy of 0.01mg.

When the adhesive mass was bound, the turbocharger rotor was mounted in the core, which was subjected to strength tests on a device allowing to accelerate the rotating element to the speed of more than 200 000 1/min.

The above method has been used for turbocharger number 038145702E, which is used in vehicles of Volkswagen brand with engine capacity of 1900cm3 and after removal of main components of turbocharger was verified its rotor in terms of balance.

In the first stage, the basic dimensions of the rotor, which are necessary for the test to be carried out correctly, were entered into the TB COMFORT balancing system. The following dimensions have been set:

- diameter of the compression wheel,
- height of the compression wheel,
- turbine wheel diameter,
- turbine wheel height,
- turbocharger shaft length,
- distance between turbocharger shaft and balancing sensors placed in supports.

Then the rotor of the turbocharger was rotated in order to verify its level of unbalance. The obtained value of unbalance correction was reduced by application of thermo-resistant mass with the use of insulin syringe (Fig. 12 and 13).

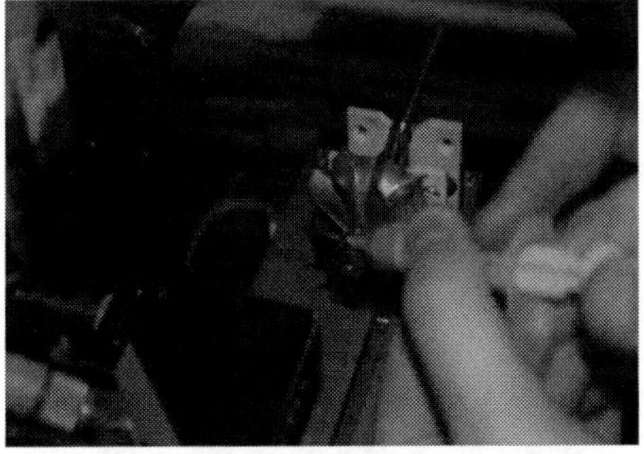

**Fig. 12.** Imposition of mass on the turbine wheel. **Fig. 13.** Imposition of mass on the turbine wheel.

After the process of balancing the turbine wheel had been completed, the compression wheel was verified. The procedure was the same as in the case of the turbine wheel. The unbalance of the compression wheel was reduced by overlapping the same mass as that of the turbine wheel. The procedure of balancing the turbocharger rotor was carried out in a correct way, the size of unbalance was within the tolerance limits.

After the process of balancing the turbocharger's rotor, the balanced element was placed in the core of the turbocharger and mounted in the high speed Schenck MBRS110 balancing machine. During this process, the rotational speed of the rotating component exceeds 200 000 rpm. At this stage of the test, the high rotational speed strength of the mass of the rotor, which is common during the daily operation of the turbocharged drive unit, has been verified. During this test, it was observed that at a rotational speed of 200 000 rpm, the heat-resistant mass remained on the turbine wheel (Fig. 14), but broke out of the compression wheel (Fig. 15).

**Fig.14.** Mass on the turbine wheel after compression test. **Fig.15.** Ripped wheel mas.

Turbochargers of properly operated car engines do not require frequent repairs and reach mileages from 100 000 to 300 000 km. They run reliably on heavy commercial vehicles, achieving mileages up to 500,000 kilometres and more. The reason of their possible earlier failures are usually mistakes in the exploitation and service of engines or damage to other systems cooperating directly with the turbochargers.

# 4 Rational use of the turbocharger

Turbocharger is an exhaust-driven compressor used for turbocharging combustion engines, which is used to increase engine power by increasing the amount of oxygen supplied to the combustion chamber - Fig. 16. The construction of the turbocharger is designed in such a way that it can cooperate with it without any problems during the engine lifetime.

The turbocharger does not need any specialist maintenance and its examination should be limited to a few basic steps included in the instructions for use of the turbocharger provided by the manufacturer. Typical damages to turbochargers are:

- damages caused by foreign material,
- Insufficient oil pressure,
- temporary delays in oil delivery,
- contaminated oil,
- overheating,
- construction faults in the process of turbocharger manufacturing.

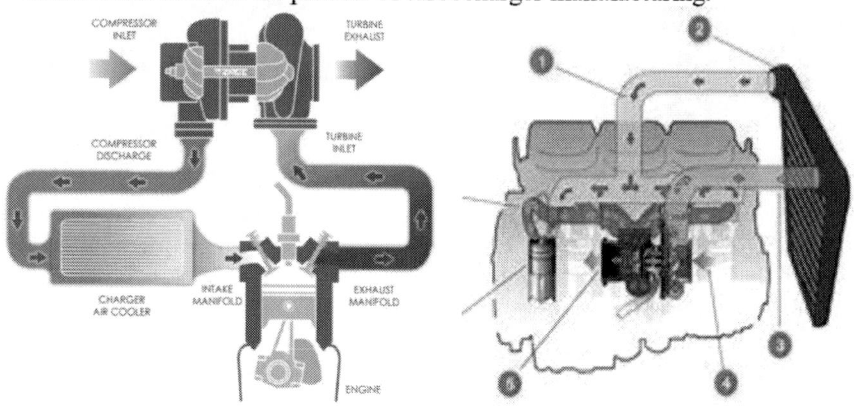


**Fig.16.** Diagram of turbo-diesel system [3].

There are obvious symptoms of turbocharger worn out:

- sudden loss of power;
- much louder operation (whistling);
- blue exhaust smoke combined with rapid oil loss in the lubrication system;
- oiling of the area around the turbocharger and in the air inlet ducts;
- („check engine") error and low engine rpm.

Approximately 90% of damages of turbochargers are caused by their incorrect operation (use and maintenance). The main causes of turbocharger damages are following:
overheating of the turbocharger (after a long period of use, it is recommended to leave the unit cooled down, lubrication decays, charring of the lubricating oil, wear of bearings, clogging of oil ducts);

- excessive load on the cold engine (cold oil is dense, the bearings are not lubricated);
- use of not recommended oil by the manufacturer (preferably synthetic, for high operating temperatures);
- exceeding the recommended oil change intervals (loss of lubricating properties);
- low oil level (bearings, tilting in the sloping);
- damaged air filter (sand, dust, high speed – wear);

- defective turbocharger control (worn actuator and valve);
- damage to the turbocharger seals (pumping oil into the engine, increase in rotation, uncontrolled damage).

Only the manufacturer of the turbocharger or authorized repair companies have the tools and personnel specialized in repairs of these devices.

The rational use of engines and turbochargers requires certain habits and simple maintenance to ensure the reliable operation of these units. The basic recommendations in this area are:

- avoiding sudden engine stop after prolonged use;
- do not use the full power of the engine immediately after it has been started up;
- use of recommended oil, resistant to high temperatures and carbonization;
- compliance with the timeliness of the oil change and the oil filter;
- regular inspection of the engine oil;
- on-time replacement of air filters;
- application of CERAMIZER (protective layer for shafts, rings, sleeves).

## 5 Summary

It is difficult to imagine a modern motor vehicle that is not equipped with an engine turbocharging system. The most common charging device is the turbocharger, which has gained its popularity thanks to its universality and flexibility. It is used both in internal combustion engines and in innovative hybrid drive units.

The most common cause of failure of turbochargers is unbalance of turbocharger's rotor. A characteristic feature of this problem is a linear increase in the amplitude of bearing vibrations, which depends on the rotational speed of the rotor. Incorrect balancing of the basic turbocharger-impeller element is caused by the following damages: cracked wheel of turbine or compression, damage to the turbocharger shaft, damage to thrust bearings and variable geometry system failure.

Analysing the proposed innovative method of balancing the rotor of turbocharger by adding heat-resistant mass we can conclude that the presented method is an interesting alternative to the current method of balancing turbochargers. Certainly, the probability of turbocharger failure during the reduction of unbalanced mass is minimized. Correction of the unbalanced mass of the rotor by the "mass addition" method does not affect the mechanical properties of the material and does not affect its structure. Thus, it does not cause cracks of rotating elements in the turbocharger.

Further research on this issue should identify the phenomenon of the loss of added mass from the compression wheel at a rotation of 200,000 min-1. In the initial analysis, this can be explained by the incorrect selection of glue for the material mass.

The proposed method aims to remove the problem of incorrect reduction of unbalanced weight. This problem appears in the form of cracks, i.e. damage to the basic components of the turbocharger's rotor. This involves expensive repairs to the main turbo-charging unit in the turbo-charging system.

The presented first results of research of the proposed solution confirm the fact that the right balance is the key aspect for the correct, failure-free use of the turbocharger.

Briefly indicated basic recommendations for rational use of car engines and turbochargers ensure safe and reliable operation of these systems.\

This paper has been achieved under the research project "Hybrid multimedia mobile scenes are a chance for decisive innovation" No. POIR.04.01.04-00-0045/17-00.

## References

1. Mysłowski J.: Pojazdy samochodowe – doładowanie silników. WKŁ, Warszawa 2011

2. Wojciechowski H. ii: Analiza metod badania turbosprężarek pojazdów samochodowych. Logistyka – nauka, nr 4/2015
3. Bojarczuk P.: Doładowanie silników spalinowych - przegląd konstrukcji i diagnostyka. Poradnik serwisowy, nr 1/2007
4. Fabryczne materiały reklamowe Mahle. Turbosprężarka na rynek wtórny
5. www.texapoland.pl
6. Materiały reklamowe Daimler Chrysler Communications. Stuttgart 2003, 2004
7. Poradnik diagnostyki samochodowej BOSCH
8. http://turbobygarrett.com
9. Merkisz J., Rychter M.: System OBD II jako przyszłe sposoby diagnozowania pojazdów. Eksploatacja i Niezawodność, nr 1, s.38-51, 2002
10. Jaskólski J., Budzik G., Marciniec A.: Balancing of turbocharger rotors. Journal of Kones 2007
11. Idzior M., Bieliński., Borowczyk T., Karpiuk W., Analiza wpływu warunków eksploatacji na stan techniczny turbosprężarek doładowanych silników spalinowych. Logistyka 2010
12. Kaszkowiak J., Borowski S., Dulcet E., Zastępowski M., Analiza uszkodzeń turbosprężarek. Logistyka 2015
13. Żółtowski B., Cempel C.: Inżynieria diagnostyki maszyn. Polskie Towarzystwo Diagnostyki Technicznej. Instytut Technologii Eksploatacji. 2004 s.1109
14. Żółtowski B.: Badania dynamiki maszyn. ATR, Bydgoszcz 2002
15. Żółtowski B.: Podstawy diagnostyki maszyn. ATR, Bydgoszcz 1996
16. Żółtowski B., Łukasiewicz M.: Wibroakustyka maszyn w laboratorium. ATR, Bydgoszcz 2005
17. Żółtowski M.: Informatyczne systemy zarządzania w inżynierii produkcji. ITE - PIB, Radom 2011

# Diagnostics of Sulfur Cement Properties

*Mariusz* Żółtowski[1,*] *Katarzyna* Jeleniewicz[1]

[1]Warsaw University of Life Sciences – SGGW, 02-787 Nowoursynowska 166, Poland.

**Abstract.** In the presented research work the subject of investigation is the construction material under the name sulfur concrete (abbreviated as sulfur concrete). The paper presents a basic description of the material which is sulfur concrete, its composition and construction, the method of production and basic mechanical properties.

## 1 Intorduction

A multitude of innovative design solutions has forced progress and creation of new construction technologies while using innovative materials. This is why we have at our disposal a huge amount of Portland cements, mortars and concretes. Each of them is characterized by appropriate properties adapted to the place of later use of this material [1,6,9].

The search for new materials, resistant to corrosion and useful especially for use in aggressive environments, has begun.

Sulfur-concrete may become such a material. The sulfur concrete can be described in its simplest form as a hot material prepared by adding organic materials to the heated sulfur, such as slag or sand. The new modified sulfur cement has been developed so that it can be used in construction as a corrosion-resistant material, with improved durability and strength.

The properties of sulfur concrete with other materials were compared, such as: mineral sulfur, sulfur polymer, mineral sulfur + gravel and ordinary concrete.

This provided grounds for determining the superiority of sulfur concrete over previously known building materials in terms of: resistance to solutions of organic and inorganic acids, salt solutions, resistance to salt mist, seawater, resistance to radiated radiation, energy, corpuscular ($\alpha$,$\beta$), and neutron.

The outlined problem of testing selected properties of sulfur concrete in comparison with other building materials fully indicates the interesting features of the new construction material, useful in many possible construction applications.

## 2 Sulfide as a construction material

Sulfur is one of the most widespread elements, with a total of thirteenth resources on earth. Particularly large sulfur resources occur on the territory of the United States, which is why they are called sulfur tycoon. In the past, sulfur was the largest source of sulfur, but their reserves were decreasing. Currently, it is produced from various sources such as: recovery from acid gas, refining, sulfur ores and recovery from coal.

Acid electrolytes and rinsing resolve the caustic attack on PVC floors, column support, but these methods are not very effective. Materials such as potassium and sodium were produced, which supported the structure without worsening the damage. That is why the search for a material that could prevent corrosion and be a very good structural material, especially used in aggressive environments, was started.

* Corresponding author: mariusz_zoltowski@sggw.pl

The sulfur concrete can be described in its simplest form as a hot-prepared material by adding organic materials such as slag or sand to heated sulfur. The new modified sulfur cement has been developed so that it can be used in the construction industry as an anti-corrosive material, with improved durability and strength.

So, we started testing a new product on a large scale. Where the new material was used, corrosion repellency could be observed, both in acidic and alkaline conditions, while maintaining its strength properties. Confirmation of this theory were the positive results of tests carried out by the American Sulfur Institute. At the beginning of 1977, cooperating industrial plants initiated a control program testing sulfur concrete in corrosive industrial environments. The main goal of the program was to establish the principles of feasibility and usefulness of the new material on a large scale of applications and determine the final properties of the material in a corrosive environment. The results were to show the useful potential of the material in corrosive environments where other materials were destroyed. An example here may be: corrosive electrolytes, acids, salts.

It was this technology of sulfur-concrete product that came to Poland. Hitherto used technologies of hazardous waste disposal consisted in neutralization and chemical treatment, followed by stabilization in various types of binders: Portland cement, pozzolanic cement, binding materials, paraffins and synthetic organic polymers, with Portland cement being the most commonly used for this purpose.

Not all hazardous waste is efficient. New, better solutions are constantly being sought [3,4,5].

The technology over which OBR PS "Siarkopol" works together with "Ekostrad" Kielce concerns the use of polymeric sulfur - sulfur cement as a material stabilizing material in the physical process, containing heavy metals, especially mercury and arsenic. Solid waste cemented with sulfur cement can be safely stored or used as elements of insulation barriers and sulfur concrete constructions, among others for construction of landfills.

These studies have confirmed that sulfur cement is better for this purpose than other binders. The stabilized waste obtained in this way is characterized by generally very low leachability of harmful components as well as mechanical strength and chemical durability, making it possible to make construction elements of them for special purposes.

## 3 Properties and application of sulfurness

Sulfur concrete as a material was developed in the seventies and eighties of the last century. Its constituent elements are: sulfur and aggregate. Plain concrete consists of cement, water and aggregate. The binding components are cement and water. In the case of sulfur concrete, sulfur is the binding element. The sulfur is chemically modified, which results in very good binding properties. No cement or water is needed. In order to obtain sulfur concrete, the sulfur and aggregates should be combined with each other. This happens at a temperature of about 140 degrees Celsius [7,8].

The binding process of this material occurs as a result of the mixture cooling. This concrete is very similar to ordinary, but its properties are slightly different. Sulfur itself is a by-product in many processes, mainly refining fuels, gases and coal.

The most important property of sulfur concrete is very high resistance to chemical corrosion. It is resistant to acids, salts and bases, as well as petroleum oils, solvents and other aggressive substances. Another important property is high mechanical strength and abrasion resistance. It is much higher than ordinary concrete. The structure of sulfur concrete has a very low porosity, and due to its hydrophobic properties, the material contained in the concrete does not rinse.

These properties result from the fact that the hardening of sulfur concrete is a physical process, unlike ordinary concrete, in the case of which we are dealing with a chemical process. The first feature is the high hardening rate, which lasts 24 hours, although in practice it takes 3-4 hours. Another property is the possibility of recycling. At the end of the useful life, the sulfur concrete can be crushed, melted and reused.

Presenting experience related to the practical application of this material, it should be mentioned that it is used in the United States, Canada and South America. Sulfur concrete has already gained widespread use there. The first of them are plants related to the processing of metal ores. In copper and zinc processing plants we deal with a very aggressive acidic environment [1,8].

The second area of application are floors in plants that deal with a highly acidic environment. Examples are fertilizer manufacturers and battery companies. In South America, sulfur concrete is used as a prefabricated element. It can be used to produce tanks for chemical products. In addition, it can be used in the production of sewerage and drainage pipes.

Very good experience gained from those countries guarantee equally effective use of sulfur concrete in Europe, eg the production of precast concrete elements has the advantage over using traditional concrete that allows you to quickly obtain full properties. Some elements can be used in road engineering as barriers or as structural elements. An additional possibility created by the production of sulfur concrete is the utilization of waste. Waste from a variety of industrial activities can be used as fillers. Examples of waste that can be used are metal waste, asbestos and organic waste. Remember that sulfur itself is also waste.

The last way to use this type of concrete is to use it for road construction in countries with very cold climates where there are problems with the use of asphalt and ordinary concrete. Sulfur concrete does not contain water. The possibilities of using this material are therefore huge. It can be assumed that also in our country, sulfur concrete will soon find the right application.

## 4 Description of modified sulfur cement

The preparation of the resisting corrosion of sulfur concrete has been changed along with the development of modified sulfur cements that have overcome the problems of instability and corrosion. Prepared dicyclopcntadiene and cyclopentadiene in a sealed oven from 6 to 12 hours at 145 ° C. The heated sulfur acquires a liquid consistency in which it can be used.

The unmodified sulfur, cooling from the liquid state, crystallizes into a homogeneous crystalline form and, cooling below 95.5 ° C, takes on a more dense, lighter and smaller volume form. After complete cooling, the sulfur cement does not undergo immediate transformation. The resulting sulfur concrete is more durable.

Modification of sulfur cement is patented by the American Trade Department, and patent rights are held by the Chemical Company of Houston in Texas, which is precisely introducing cement on the market [8].

Modified sulfur concrete is a hot-prepared material, mixing cement and admixtures at a temperature of 125 ° C to 150 ° C. The resulting product is cast into a form giving the desired shape and then it is cooled. This process is a thermal process that can be repeated when necessary.

In construction, more and more often a good quality construction material containing components for combating corrosion is required. The sulfur concrete must be clean, have low absorption, and dried should contain less than 0.25% water. If corrosion is caused by salt, quartz, limestone, we use rounded gravel. Rough surfaces give a better bond for a sulfur binder than polished surfaces of rounded gravels.

The cement requirements for the preparation of sulfur concrete are such that when we use more sulfur, we add less binder, and the mechanical properties remain the same. The mix should be designed according to the aggressiveness of the environment in which it will be used.

## 5 Industrial use of sulfur concrete

Sulfur concrete is prepared hot by mixing modified sulfur cement and additives. It is recommended to use a mixer that will keep the mixture at the right temperature.

Testing the quality of sulfur concrete consists in inserting ready samples into aggressive environments together with Portland cement and comparing the results.

The sulfur concrete is poured hot and must be cooled to the point where it begins to solidify. Under normal conditions, it takes approximately 30 minutes. Insulated tanks can be used to transport sulfur concrete from the mixer to the casting. A vibrating probe is used to densify the sulfur mixture, especially when used in large quantities.

Sulfur concrete is used in sulfuric acid and ammonium sulphate plants in sewers and sumps (Fig. 1), where despite contact with hot sulfuric acid (concentration up to 98%) there are no problems. It was used both as a monolith and as a protective coating on existing concrete.

**Fig. 1.** Sulfur concrete used in the construction of sewers and drains.

Sulfur concrete floors have also been installed in many rooms with electrolytic baths, showing exceptional functionality there (Fig. 2).

**Fig. 2.** Floors made of sulfur concrete.

Field tests with the performance of road pavements have been successfully carried out using traditional machines for asphalt pavement construction, equipped with heated, vibrating leveling boards (no need to roll for compaction) and standard machines for building concrete pavements with pre-heating of leveling boards. Road elements, such as central barriers, road signs and traffic lights, curbs, railroad crossings - all of this type of sulfur concrete products are successfully used in Calgary. Such elements are shown in Fig.3.

**Fig. 3.** Road elements, such as central barriers made of sulfur concrete.

Runways and airport plates, attempts to repair airport plates using sulfur concrete were very encouraging - quick consolidation is a very important factor here.

Floors in food processing plants are exposed to highly corrosive conditions, not only organic acids that can soak into concrete surfaces, but also cleaning and disinfecting agents used daily for sanitary reasons. In 1987, concrete was approved for use in food processing plants in both Canada and the United States. Inspectors from Agriculture Canada encouraged its use to solve the corrosion problems that occur with ordinary concrete and concrete-liner combinations.

## 6 Sulfur production in Poland

In Poland, MARBET in Tarnobrzeg has recently been engaged in the production of sulfur concrete. The obtained sulfur concrete (SULTECH®) allows stabilization of a number of hazardous wastes containing barium, zinc, chromium, mercury, heavy metals and asbestos (confirmed by tests of research units) - Table 1.

The resulting sulfur concrete containing stabilized asbestos is characterized by very good parameters (confirmed by ITB, PZH), such as:

- high strength of products;
- resistance to acids, seawater, sewage;
- high frost resistance;
- lack of absorbability;
- no electrical conductivity (perfect dielectric);
- short time of forming and obtaining nominal strength;
- the possibility of recycling products (thermoplastic).

In the years 1999-2001, many industrial applications using sulfur polymer were developed at MARBET. It was also possible to use SULCEM® polymer for stabilization of waste. This process is particularly useful in the case of stabilization of waste containing heavy metals such as cadmium, chromium, arsenic, barium salts, salts of other metals, ashes from waste incineration plants, as well as asbestos-containing elements. Past effects.

We offer our own original technical and technological solutions proprietary to the process of recovery of hazardous solid industrial waste by stabilization using sulfur polymer in SULBET® polymer concrete.

**Table 1.** Characteristics of sulfur cement.

| COMPARISON OF PHYSICAL AND MECHANICAL PROPERTIES OF SULFUR CONCRETE AND TRADICIONAL CONCRETE | | | |
|---|---|---|---|
| Lp. | **Properties** | **Sulfur concrete** | **Tradicional concrete** |
| 1. | Density | 2.4 g /cm$^3$ | 2.2 g /cm$^3$ |
| 2. | Compressive strenght | 60 - 115 MPa | 15 - 60 MPa |
| 3. | Flexural strenght | 10 – 16 MPa | 6 - 7 MPa |
| 4. | Young's module | 40 - 65 GPa | 25 - 28 GPa |
| 5. | Shrinkage | 0.5 - 1.0 mm /m * | 0,6 mm/m |
| 6. | Thermal expansion | 10 - 30 $\cdot$ 10$^{-6}$ 1 / °C | 8 - 10 $\cdot$ 10$^{-6}$ 1/ °C |
| 7. | Thermal conductivity | 0.4 - 2 W /m °C | 1,4 - 3,6 W /m °C |
| 9. | Impregnability | 0 – 1 % | 5% |
| 10. | Frost rezistance | F>500 | F50 |
| 11. | Time required to form an element | 0.1 - 0.4 h | 48 h |
| 12. | Resistance to aggressive environment | full | non |
| 13. | The proportion of binding material in the finished concrete | 10 – 15% | 20 - 30% |

This method, due to the good mechanical and physicochemical properties of the produced polymer concrete (high resistance to acids and salts), can be widely used in waste recovery, e.g. for the production of:
- concrete elements used for lifting shafts;
- flood protection (SULTECH SYSTEM NOE®), for reinforcing soils preventing landslides as well as for the production of pipes with chemical resistance, etc. elements embedded in the ground,
- elements of manholes and covers with finial for telecommunications,
- concrete elements for road construction: road plates, openwork plates intended for car parks, culverts and road borders, linear drainage, sewer and sediment manholes .;
- weights for flooded pipelines and cables of railway traction and weights for automatic washing machines.

The main advantages of the SULBET® polymer concrete produced include:
1. technological features:
- the possibility of recycling products in the future (thermoplastic),
- short time of concrete elements curing and reaching their nominal strength (table 2).
2. functional features:
- very good mechanical properties (high compression and bending strength) exceeding the analogous Portland concretes (ITB test results, table 2),
- resistance to aggressive media: acids, salts, oils, municipal sewage, sea water (ITB),
- excellent corrosion resistance (ITB tests),
- tightness and very high frost resistance (ITB tests, table 1),

- very low absorbability (table 1),
- a comparable coefficient of linear expansion with concrete based on Portland cement (Table 1),
- possibility of making reinforced concrete elements - SULBET® polymer concrete has a higher binding strength to reinforcing steel than Portland concrete and protects it against corrosion (ITB tests),
- it can be used to stabilize hazardous waste (eg ashes from waste incineration plants) without deteriorating its properties (ITB research).

**Table 2.** Benefits resulting from the replacement of conventional concretes with sulfur concrete.

| Infastructure | Application | Important properties | The advantages of using sulfur concrete |
|---|---|---|---|
| Sewage | sewage pipes, manholes, processing plants | strength, durability, resistance to corrosion, impermeability, not leaching | - increased durability<br>- sewage pipes do not require protective lining<br>- lower cost |
| Solid waste | tanks, sheath (hermetic) | strength, durability, resistance to corrosion, impermeability, not leaching | - increased durability<br>- longer life<br>- lower cost in terms of the utility cycle |
| Radioactive waste | burials | durability, resistance and radiopacity, | - higher resistance and impermeability of radiation than lead |
| Transport | bridges, road surfaces | durability, impermeability, quick consolidation, durability | - increased durability<br>- quick consolidation |
| Agriculture | animal pens, sewers, troughs | resistance to corrosion, impermeability, strength, durability | - hygiene improvement<br>- lower cost<br>- ease of cleaning |
| Chemical Works | floors, sumps, tanks, sewers | corrosion resistance, imperviousness strength | - durability<br>- quick consolidation |
| Food industry | floors, walls | resistance to corrosion, impermeability, durability | - ease of cleaning<br>- quick consolidation<br>- shortens breaks at work |
| Marine environment | breakwaters, buoys, docks, wharfs for barges | corrosion resistance, durability, impermeability, durability | - durability<br>- ease of repair or renewal |
| Water | pipes, tanks, pools | corrosion resistance, durability, impermeability | - durability<br>- longer life<br>- lower cost |

## 7 Use of sulfur concrete for road repair

The subject of the invention is a method for producing mixtures based on a sulfur binder for road repairs, used as a construction material with increased resistance to dynamic loads, useful especially in difficult weather conditions.

The method for producing sulfur-based binder mixes for road repairs containing sulfur and / or sulfuric polymer is characterized in that the sulfur and / or sulfur polymer is heated until it reaches a liquid or semi-liquid state, after which it is sprayed, preferably under the effect of the overpressure, whereby mineral aggregate and metal filler in the form of preferably short metal fibers are introduced into the so formed mist for the time of complete coating of the mineral aggregate and the metal filler with sulfur and / or sulfur polymer, and then cooled to ambient temperature.

## 8 Summary

The content of this work discusses selected issues related to the new building material, which is sulfur concrete. The outlined problem of testing selected properties of sulfur concrete in comparison with other building materials fully indicates the interesting features of the new construction material, useful in many possible construction applications.

Even cautious inference supported by the study of this work indicates that the use of sulfur concrete in environments exposed to corrosion is technically and technologically feasible. This material shows excellent mechanical properties, much better than other materials, e.g. Portland cement.

A certain obstacle in the use of sulfur concrete, however, is the flammability of sulfur and its low melting point. Under the influence of fire, the sulfur binder in the concrete burns (however, immediately after removing the fire it goes off immediately). Sulfur concrete also softens when heated, which is why it is not suitable for constructing structures working under load.

## References

1. I. Biczok, *Concrete corrosion and concrete protection* (Budapest, 1964)
2. R.T. Conley, M. Dekker, *Thermal stability of polymers* (New York 1970)
3. F. Lea, *The chemistry of cement and concrete* (London, 1970)
4. W. Lucas, *Cement and concrete research* FCD **Vol. 5**, (1975)
5. M.V. Moskvin, *Korozja betonu* (Moscow 1952)
6. R. Shalon, M. Raphael, *Sulfur concrete proprieties* JACI **Vol. 55** (1955)
7. T. Sikorski, *Podstawy chemii i technologii polimerów* **PWN** (Warszawa 1985)
8. W. Schnabel, *Polymer degradation* **Akademie-Verlag** (Berlin 1981)
9. M. Żółtowski, *Właściwości betonu siarkowego w środowisku zasadowym* **UWM** (Olsztyn 2004)

# AUTHOR INDEX

**EDP Sciences – Web of Conferences**
17, Avenue du Hoggar
Parc d'Activité de Courtabouf
BP 112
F-91944 Les Ulis Cedex A
France

ISBN 978-1-5108-9931-5

9 781510 899315

CURRAN ASSOCIATES INC.
proceedings.com

# Research, Invention, and Innovation Congress (RI2C 2022)

Held Online

E3S Web of Conferences Volume 355 (2022)

Bangkok, Thailand
4-5 August 2022

edp sciences